RIVERS FORGOTTEN

JEREMY KAI

Foreword

Since 2006, Jeremy Kai has brought a renewed energy to the task of exploring, documenting and celebrating the wastewater infrastructure beneath Toronto. His work has helped to invigorate a local community of photographers and aficionados to continue the quest to see and understand the buried, hidden infrastructures that make the landscape and life of the city possible. Our vision of the city is richer for his efforts, and his contribution continues here in this volume.

The wide ranging imagery in the pages ahead takes the viewer from North York suburbs to Eastern Beaches, through the parallel Toronto that extends beneath our parks and streets. This architecture is the intermediate destination of our rainwater, flushed toilets and emptying sink drains, the incredibly extensive spaces that connect your street to the lake, and your house to a sewage treatment plant. In many parts of Toronto's pre-war core, all of this rain and wastewater is carried in one combined pipe or tunnel—to a treatment plant when the weather is good, and directly into the watershed and lake when rain overwhelms the capacity of these combined sewers.

The last few decades have seen efforts in cities across North America to address sewage overflows and to raise citizen awareness and proactive action around wastewater issues. Yet we frequently ask urban residents to take action to address problems with infrastructure that they haven't been allowed to see, to understand and to make an honest connection within their everyday lives. The result is expert-driven decision-making and an enduring public disinterest in issues of urban drainage and wastewater handling that should be at the forefront of our conversations about the problems and future of our cities' ecology.

Photographs like Kai's, presented to the public, are the first step in rectifying this dissconnect even if they do not garner the support of disinterested public works departments. Kai's perspective offers introduction, inspiration, and a desperately needed public vision into the critical depths of our urban environment.

Michael Cook
vanishingpoint.ca

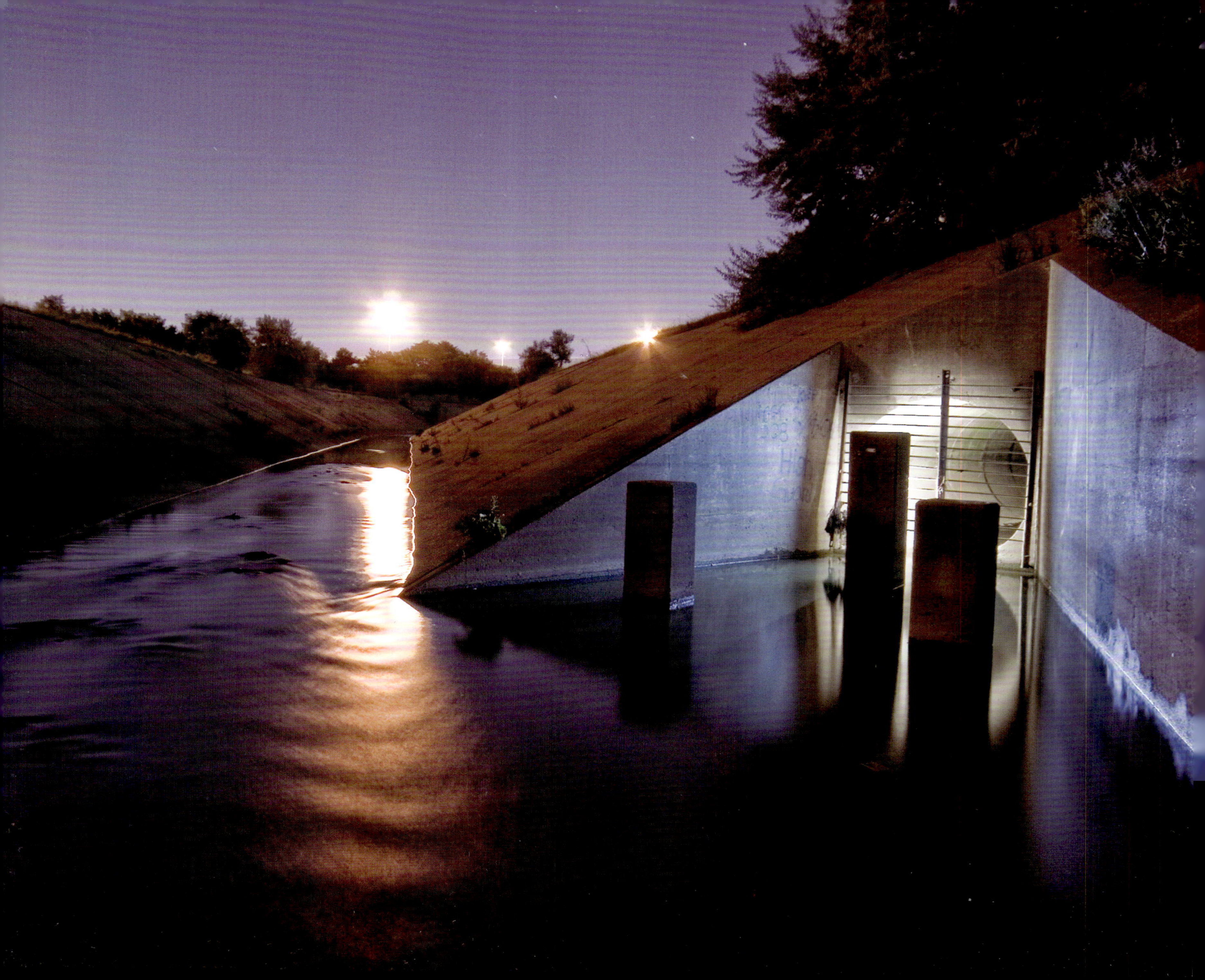

During my early adulthood, I began to feel a magnetic pull to the unknown and forgotten landscapes of the city. I used to pass over a creek on my route to high school where the water had been diverted through a concrete channel. On a sunny day in summer 2006 I hopped over the fence and walked down the steep slope where two storm drain outfalls sat at either end of the creek like gateways to other dimensions.

My first experience inside those drains enlightened me. I had only a cheap flashlight, my camera and tripod with me, along with my vague understanding of how to take pictures in the dark and the equally vague understanding of how and why drains are built.

During that first encounter, I didn't travel through the concrete tube much further than where I could still see sunlight from the outfall. In the distance I could hear a constant stream of water flowing but it seemed impossible to know how far away it was as every footstep made a distorted splashing sound that bounced at weird angles on the curved surfaces. I couldn't even think what the source of the sound would even look like. The only thing I really understood about drains was not to go while it was raining. I was both curious and apprehensive to explore more, but ultimately intrigued.

I remember the first time I heard a car run over a manhole. It sounded like a mythological blacksmith hitting an anvil with an oversized hammer. The crash echoed through the concrete pipe like nothing I ever heard before.

As I continued to explore Toronto's underground I became particularly interested in why cities require storm drains.

In cities that are defined by a significant water feature, most often a network of rivers and streams once occupied parts of the land. In the pursuit of development, the spaces that these lesser water features occupied were valuable land to build on. In order to meet our needs, we paved roads, placed sidewalks, constructed buildings, and filled in the land.

In some older parts of Toronto there are clues on ground-level to where some of the buried creeks exist. Perhaps the most mythologized lost river is the Garrison Creek that once flowed through Christie Pits, Little Italy, and Trinity Bellwoods park, south to Old Fort York.

Perhaps the best indications are the deep ravines in the parks along its former surface route. A secondary indication is something you can only really see by looking at maps. Much of the development around the creek existed before its burial so the roads were designed around it, throwing off Toronto's mostly rigid grid layout.

I came to realize that these systems were important pieces of infrastructure that now define the landscape of Toronto, even if most people don't realize it.

The city's sewers are one of the few places where you can witness generations of civil engineering that are unique to Toronto. The burial of creeks, on one hand, razed natural landscapes, but allowed for development. Eventually, the newer generations of citizens will forget that they live on top of a riverbed, within an urban watershed.

Portals

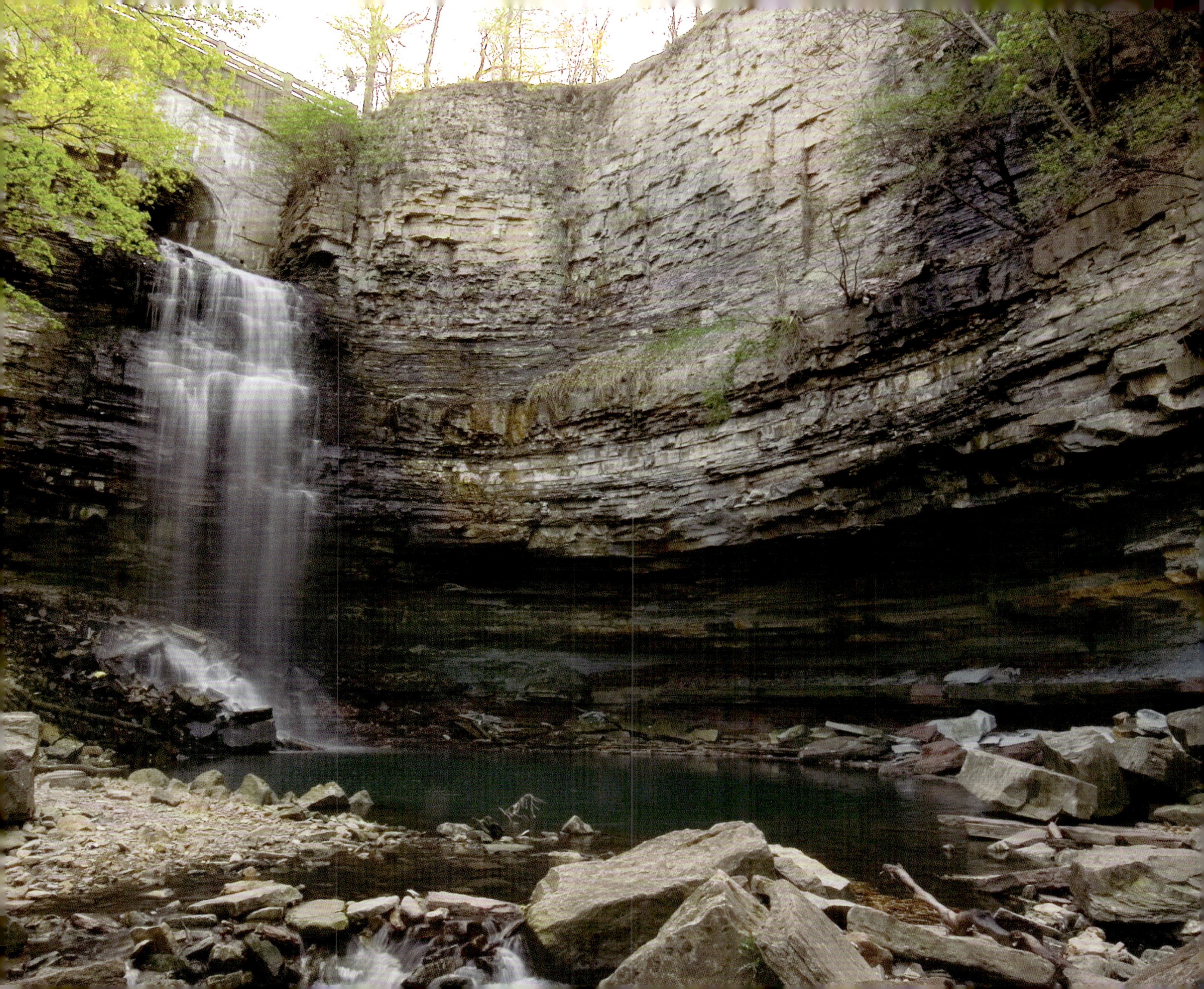

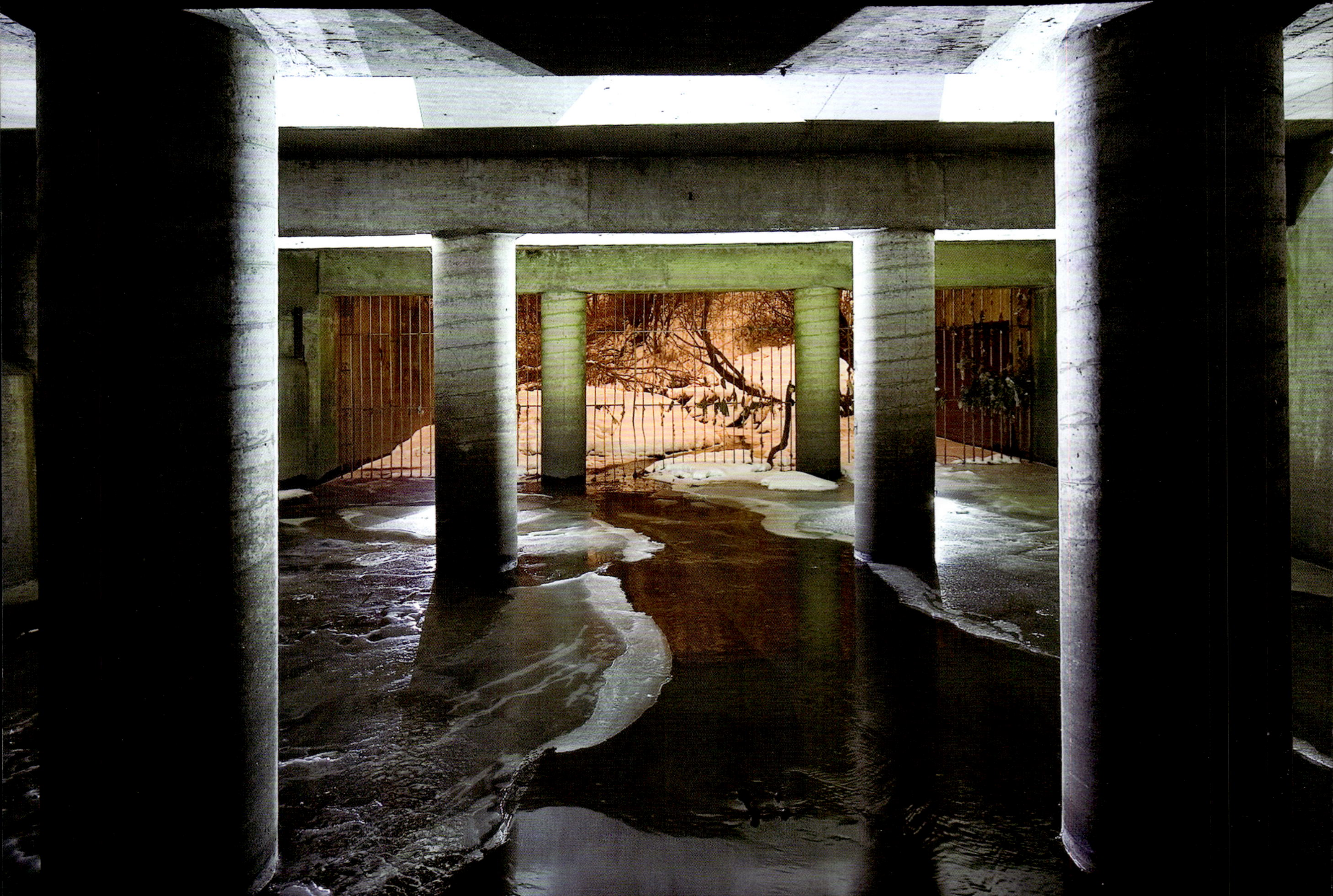

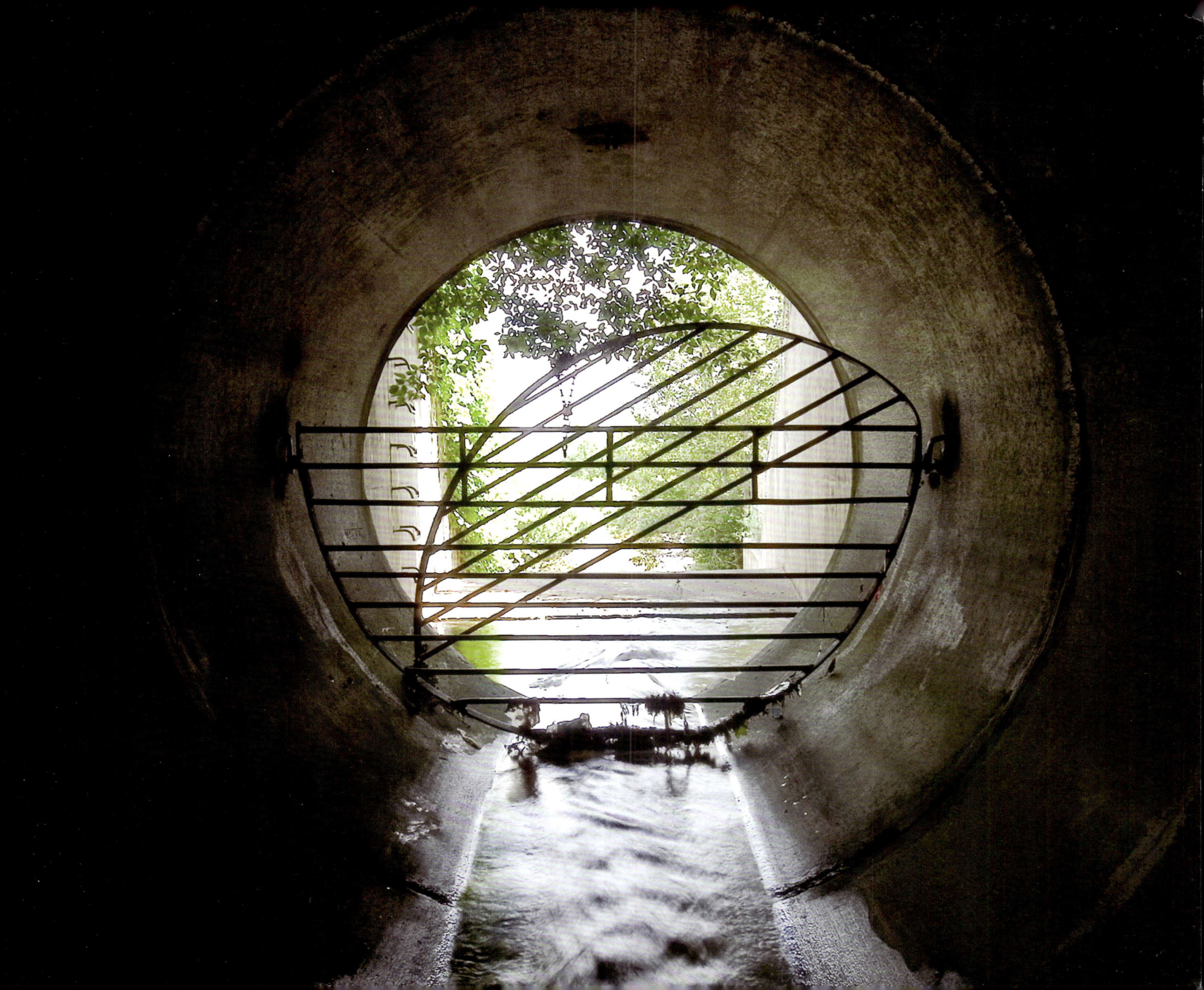

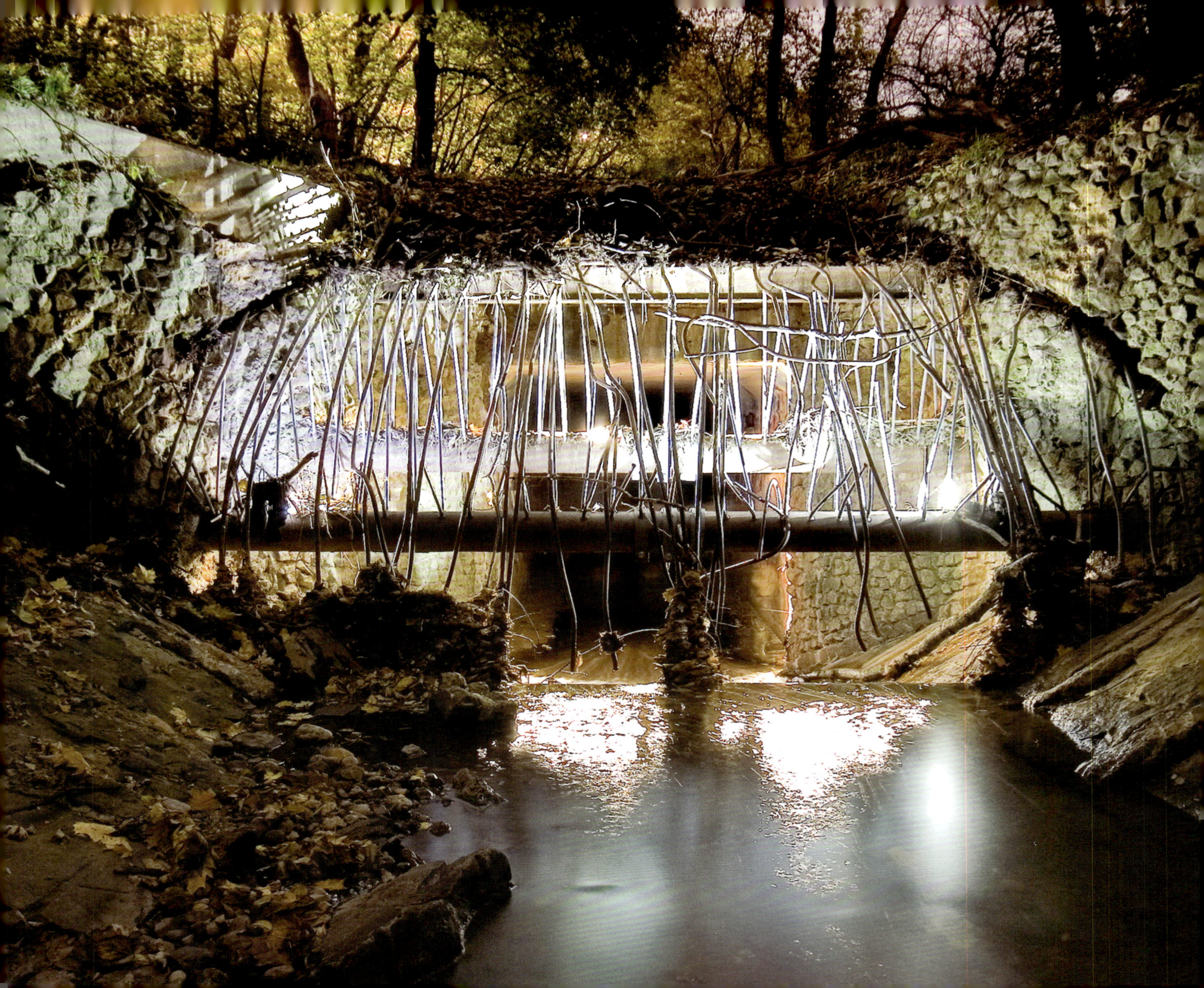

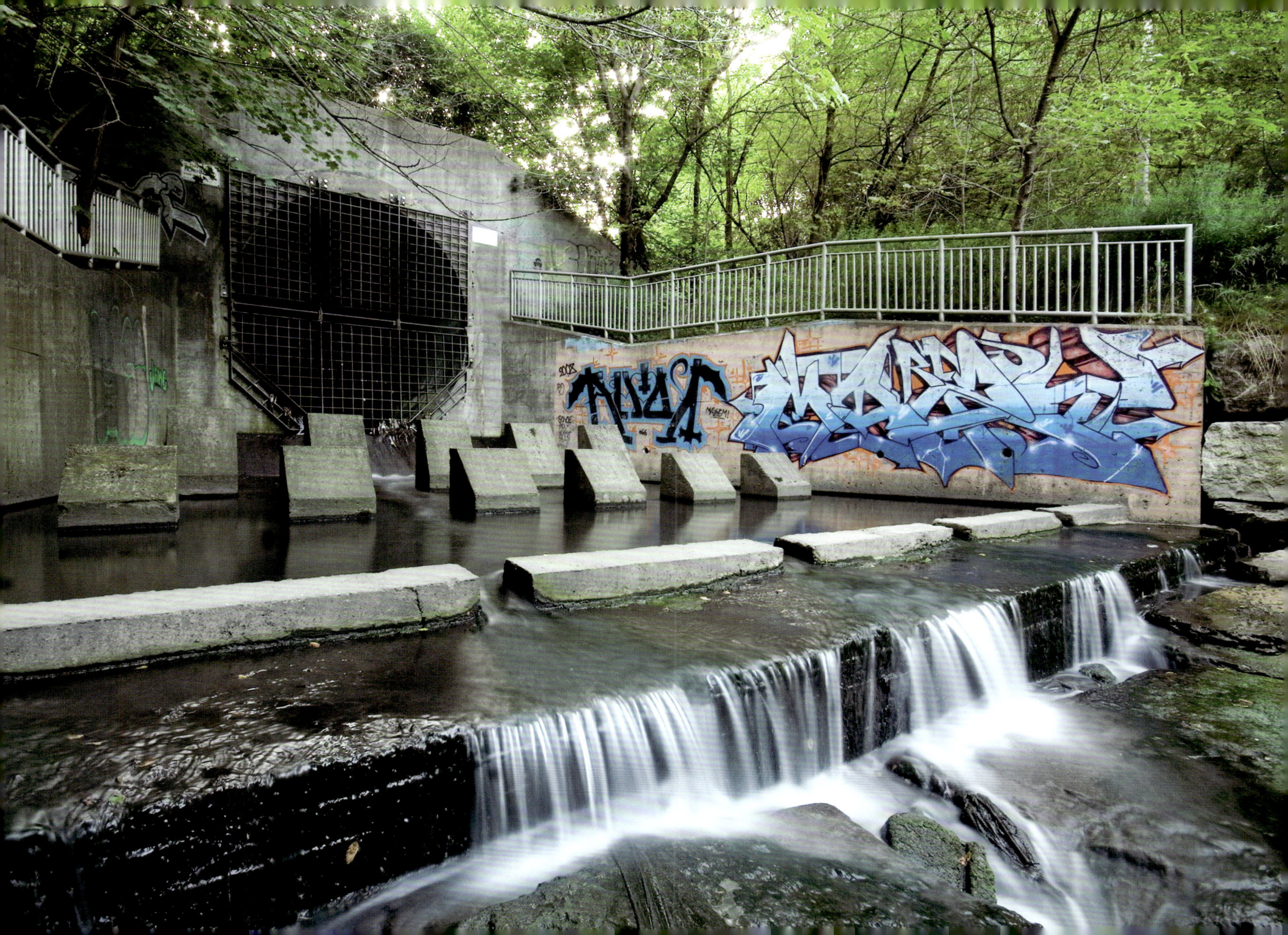

Passageways

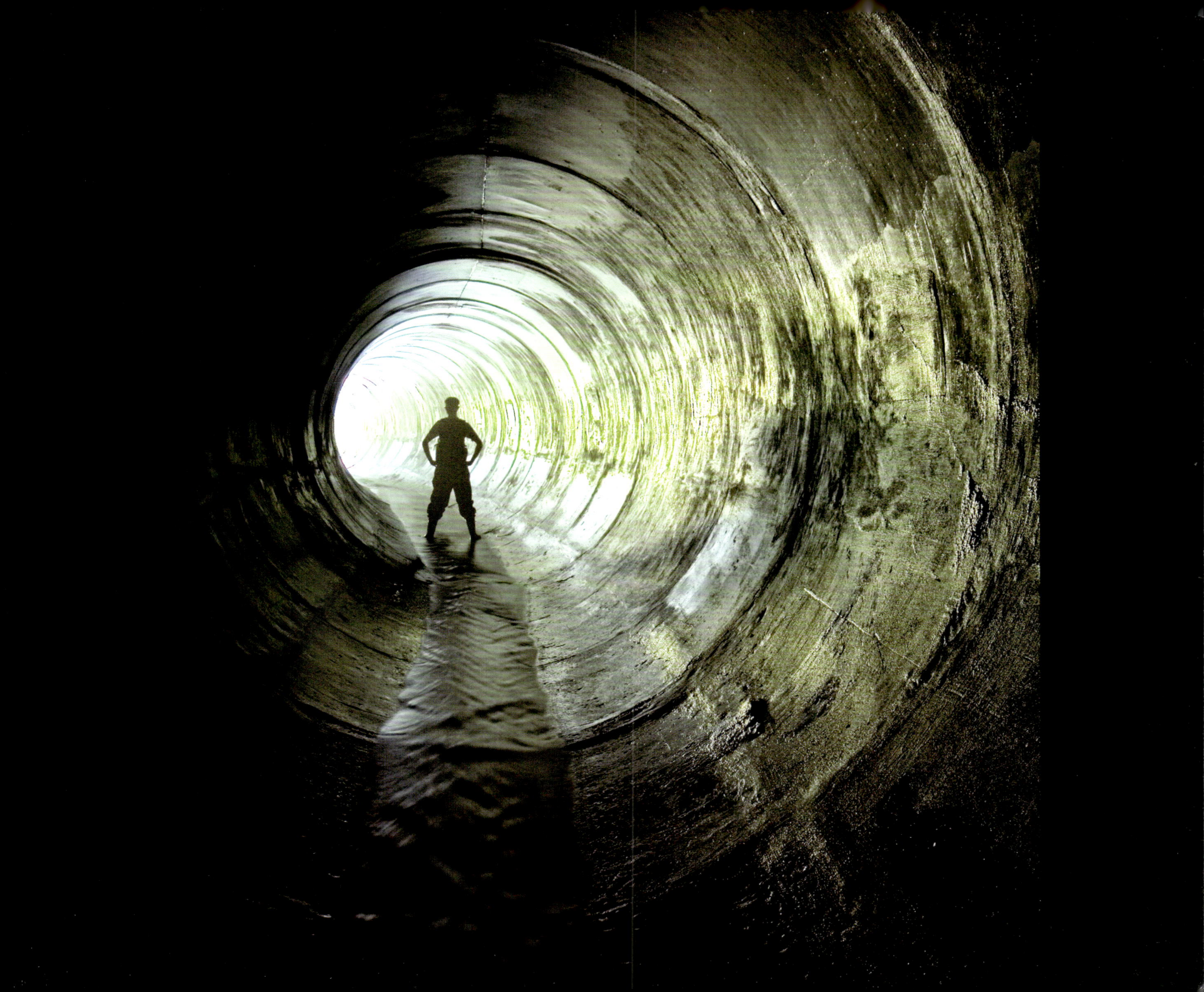

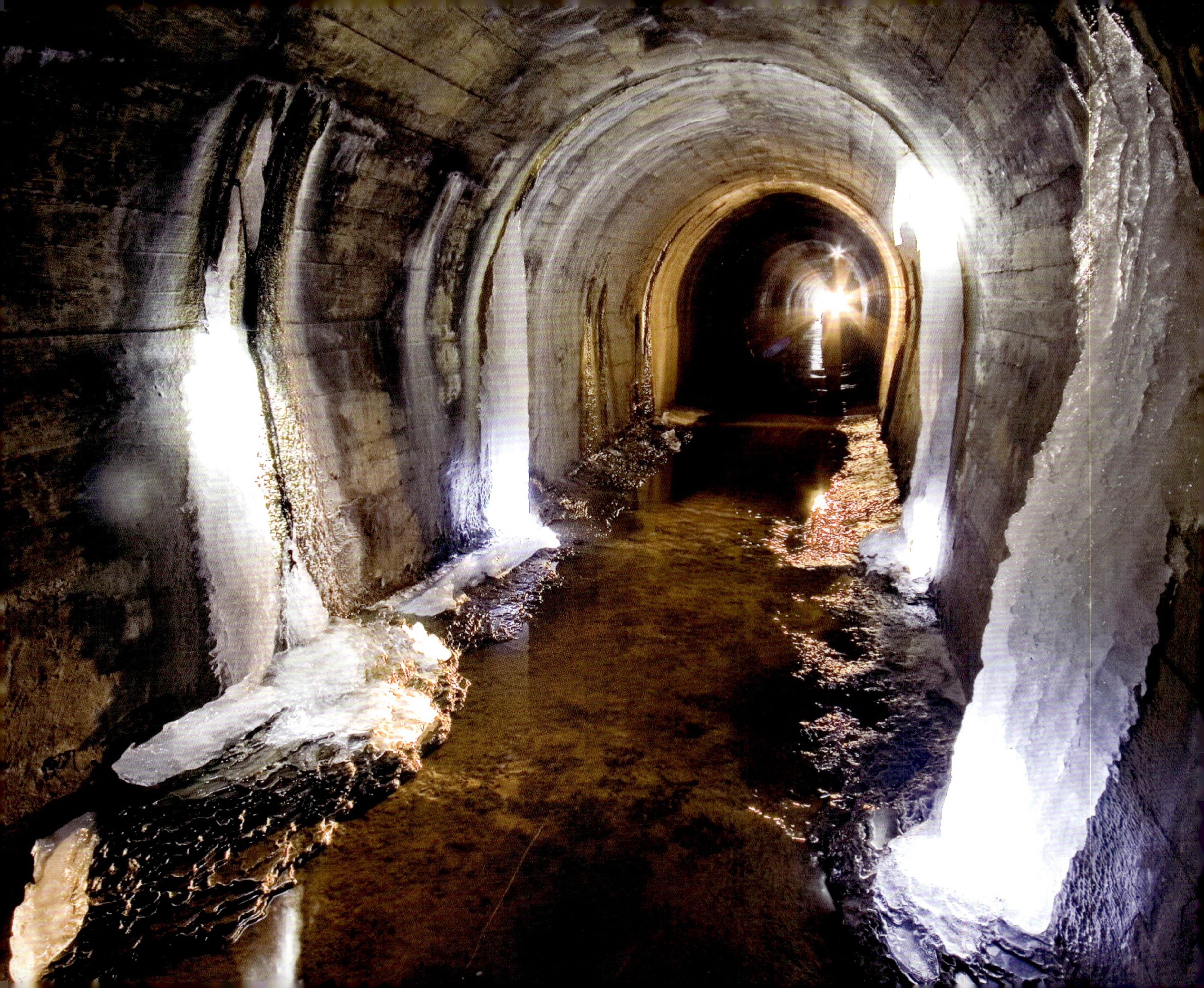

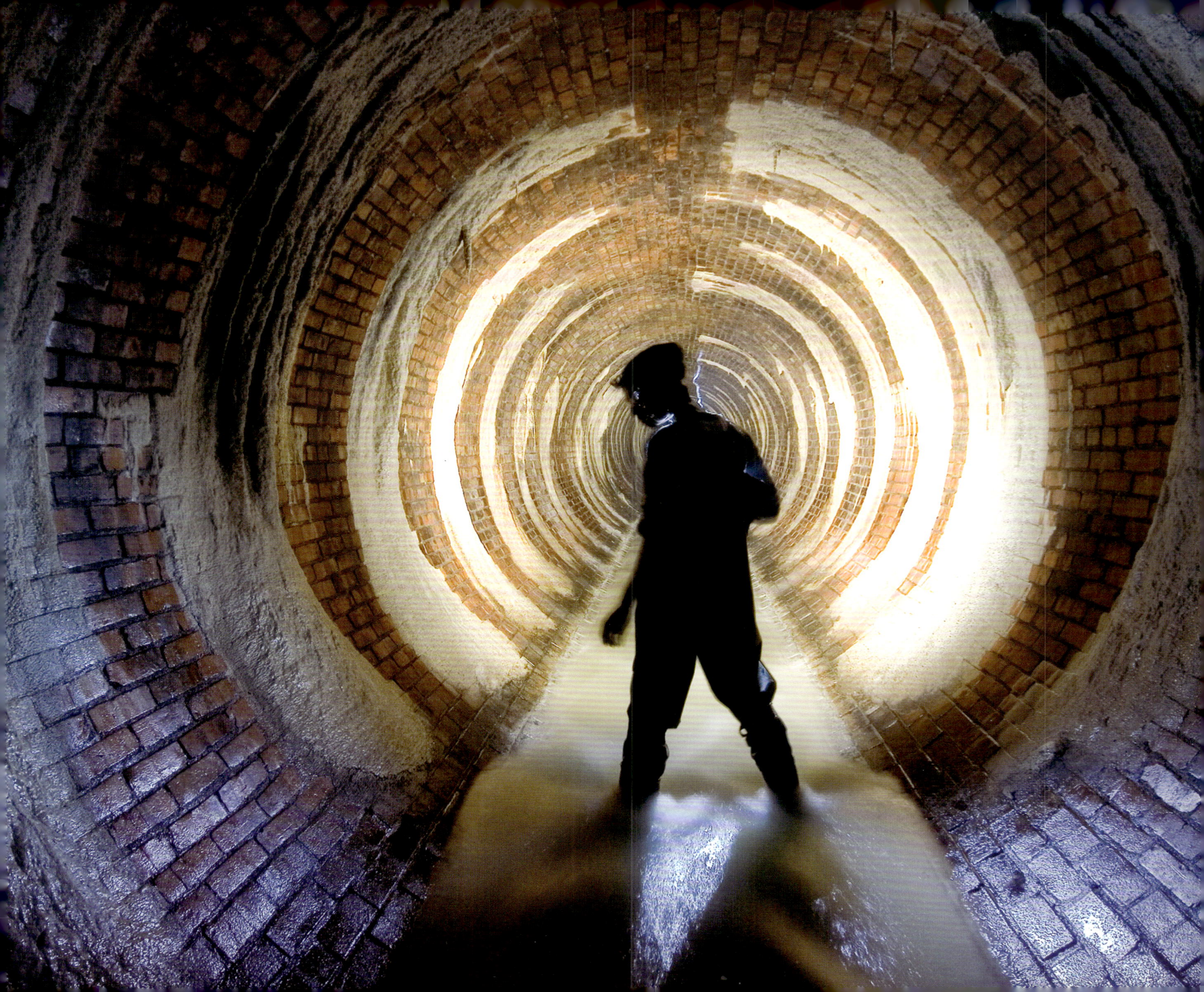

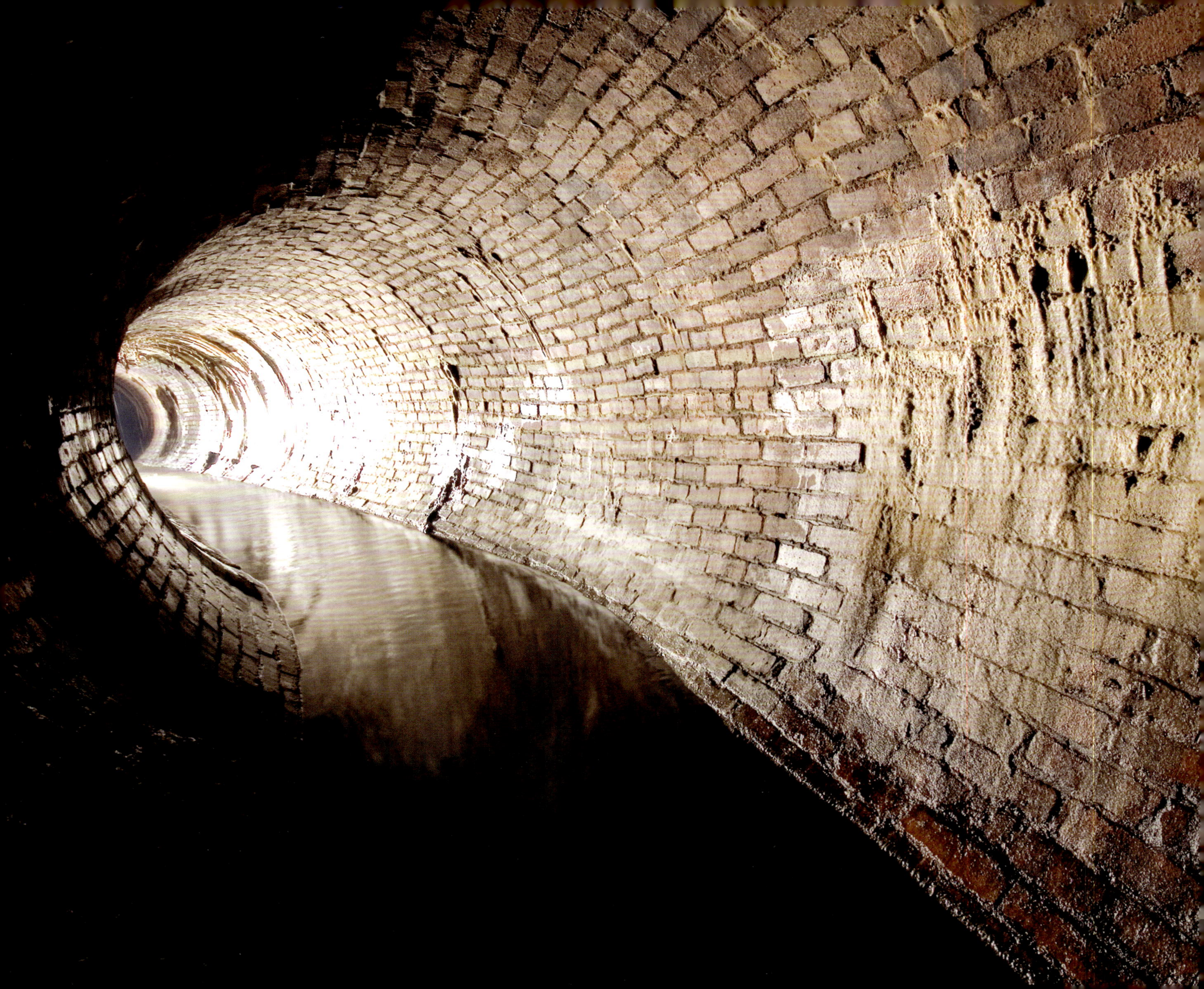

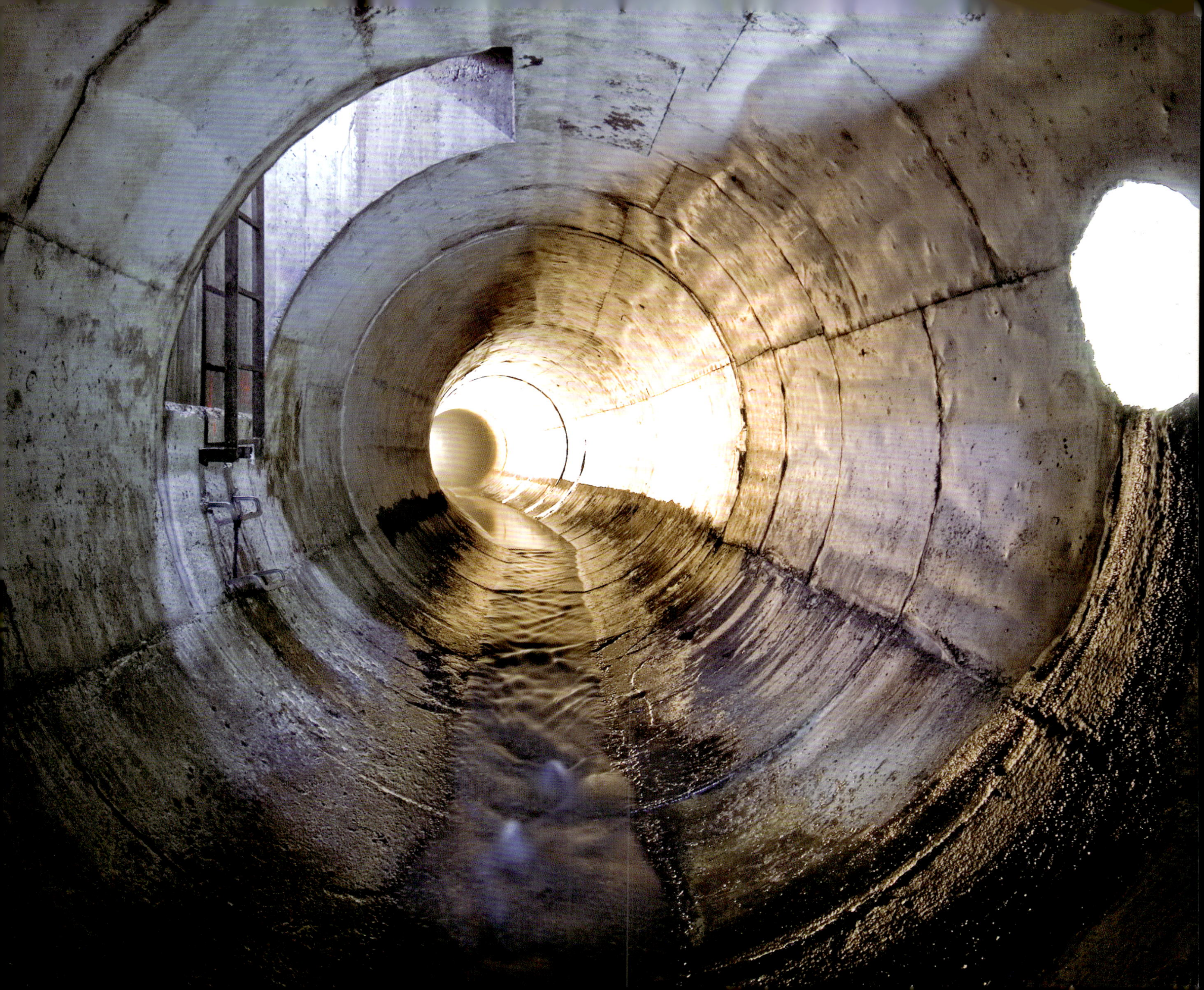

Intersections

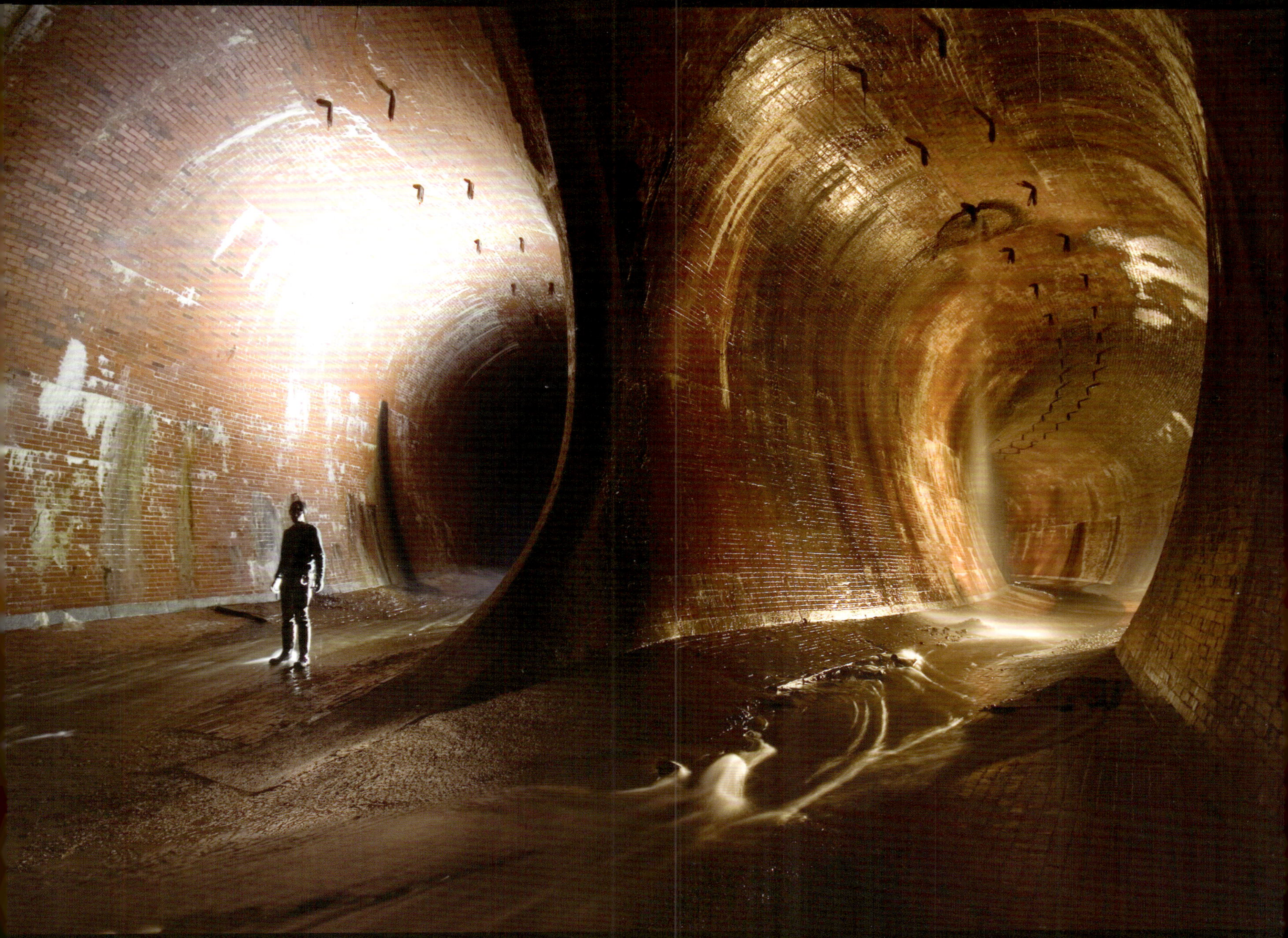

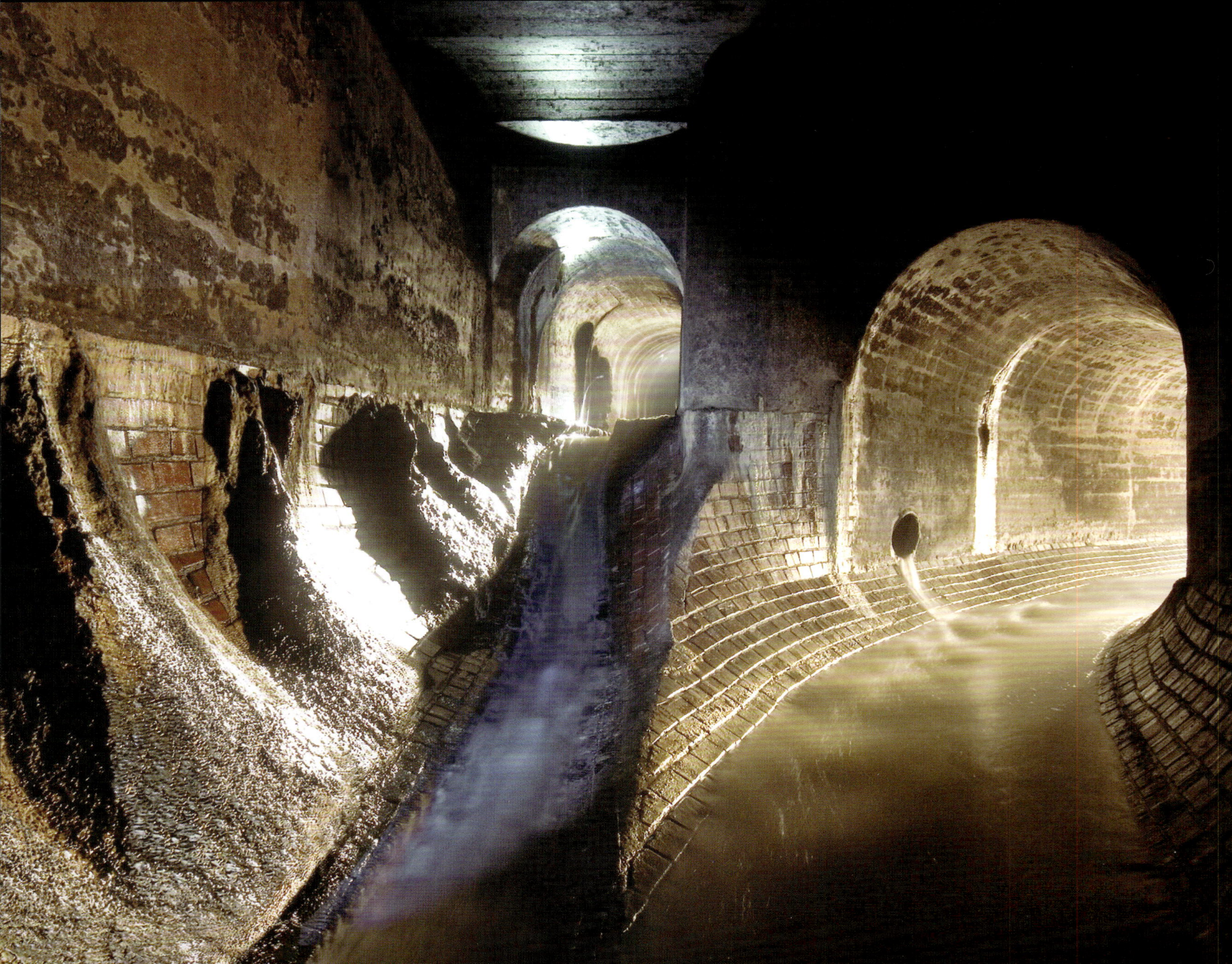

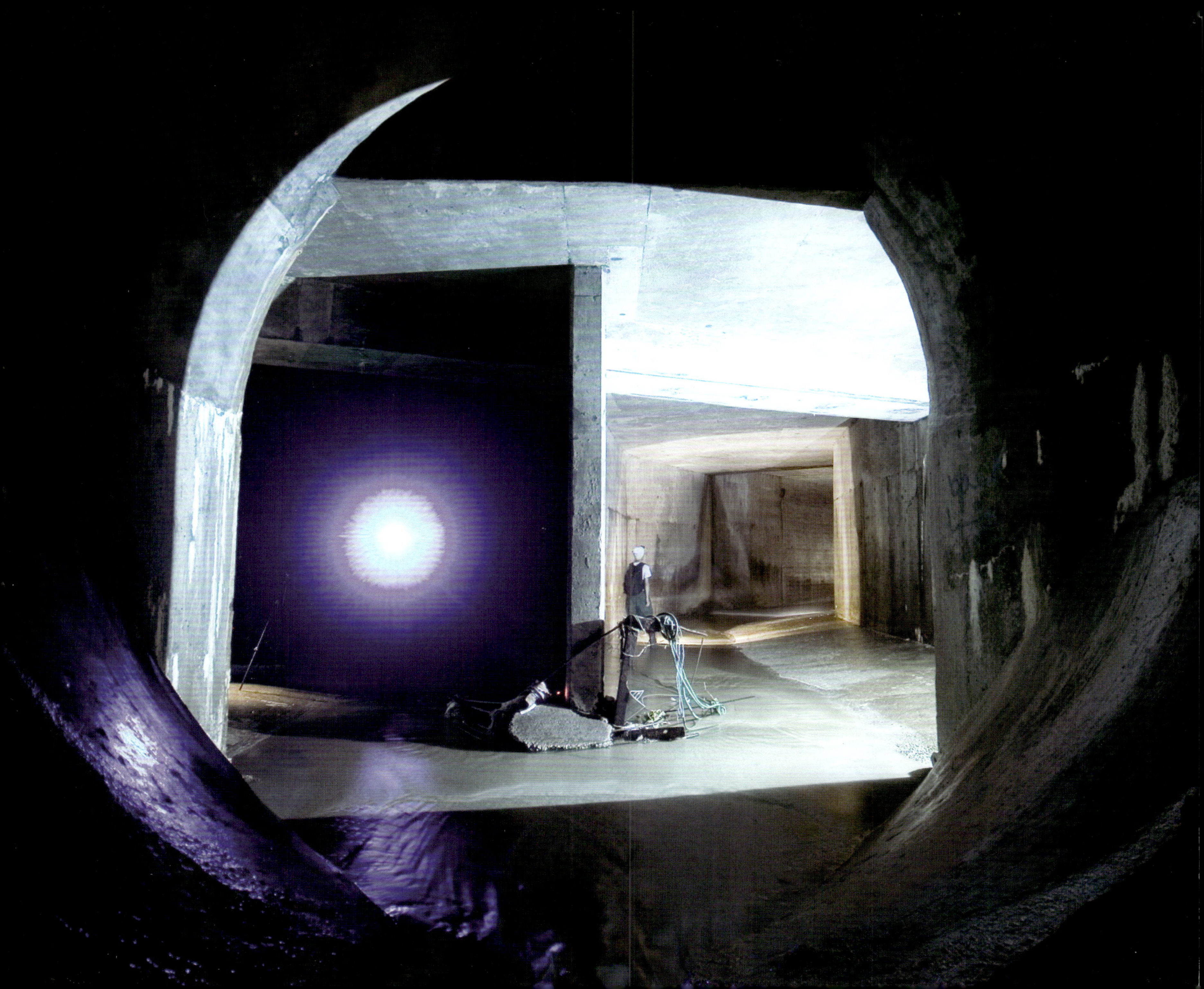

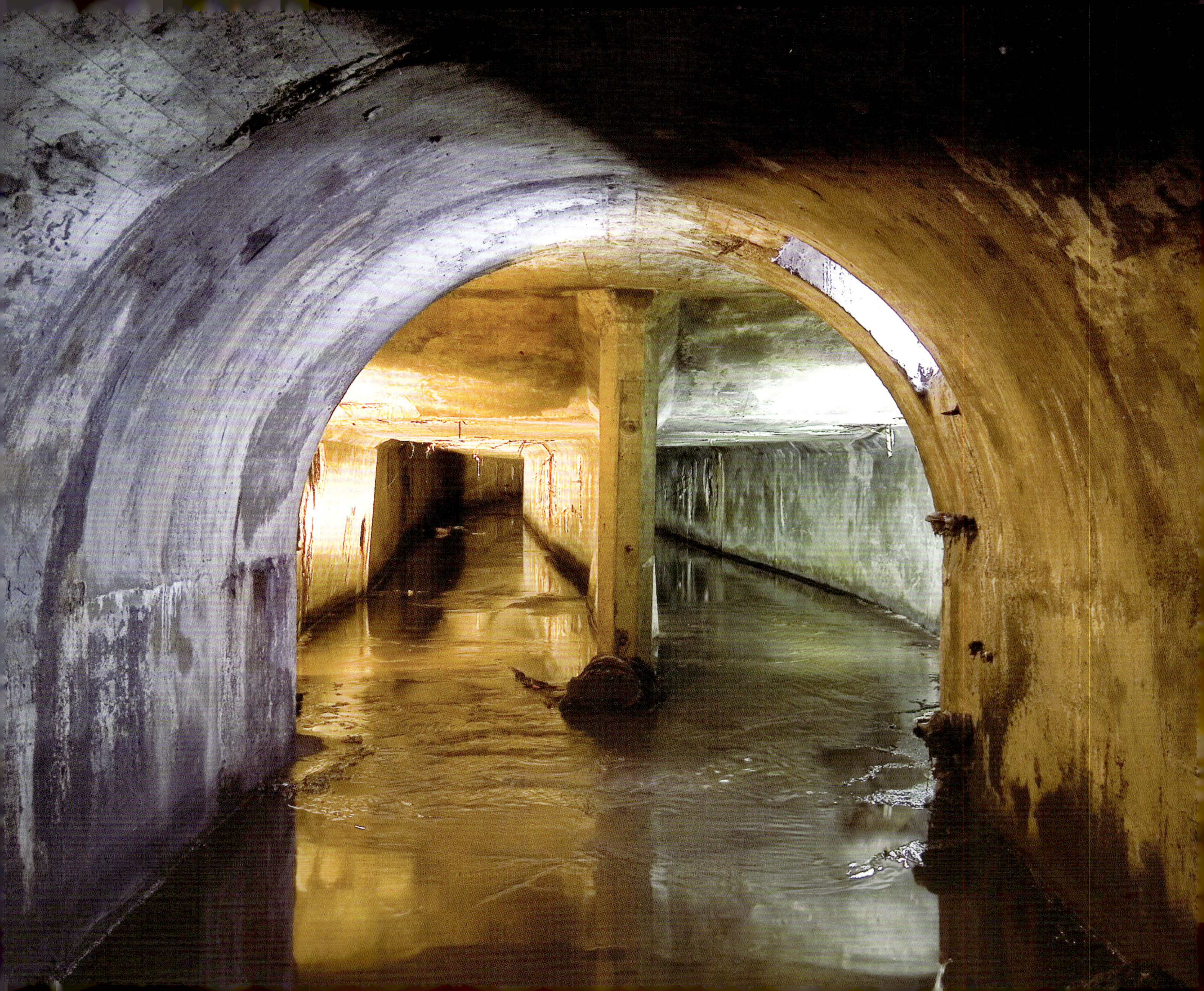

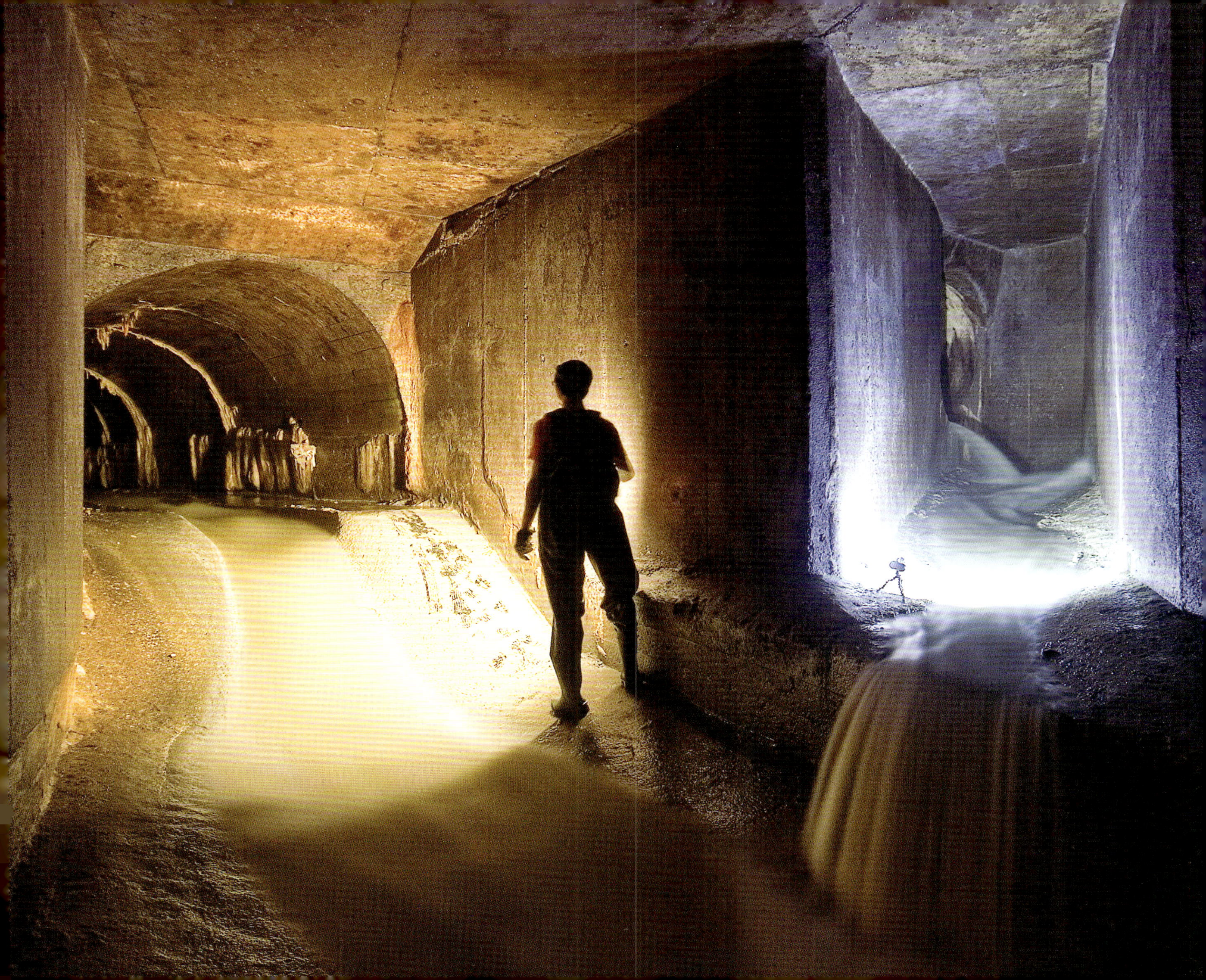

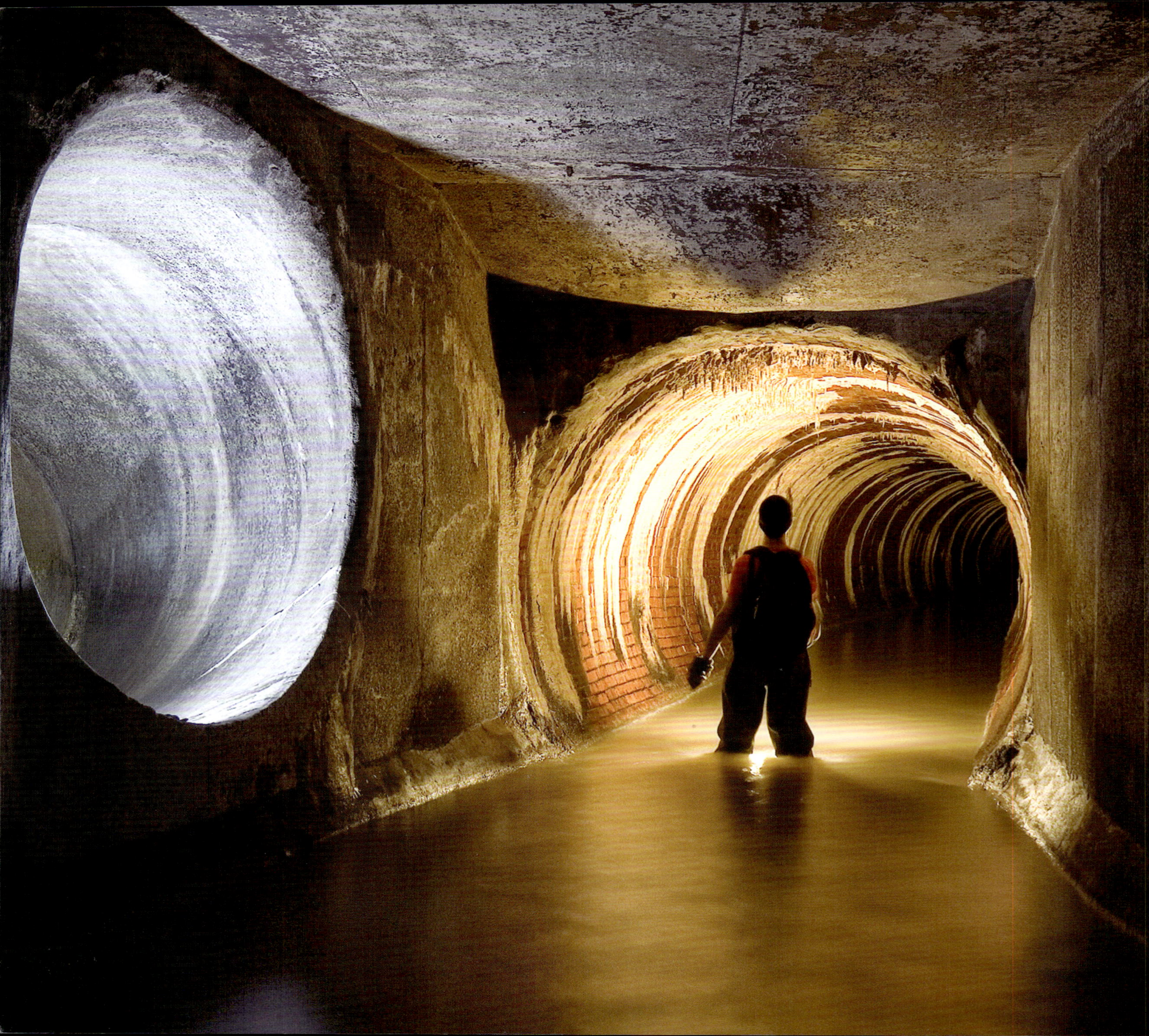

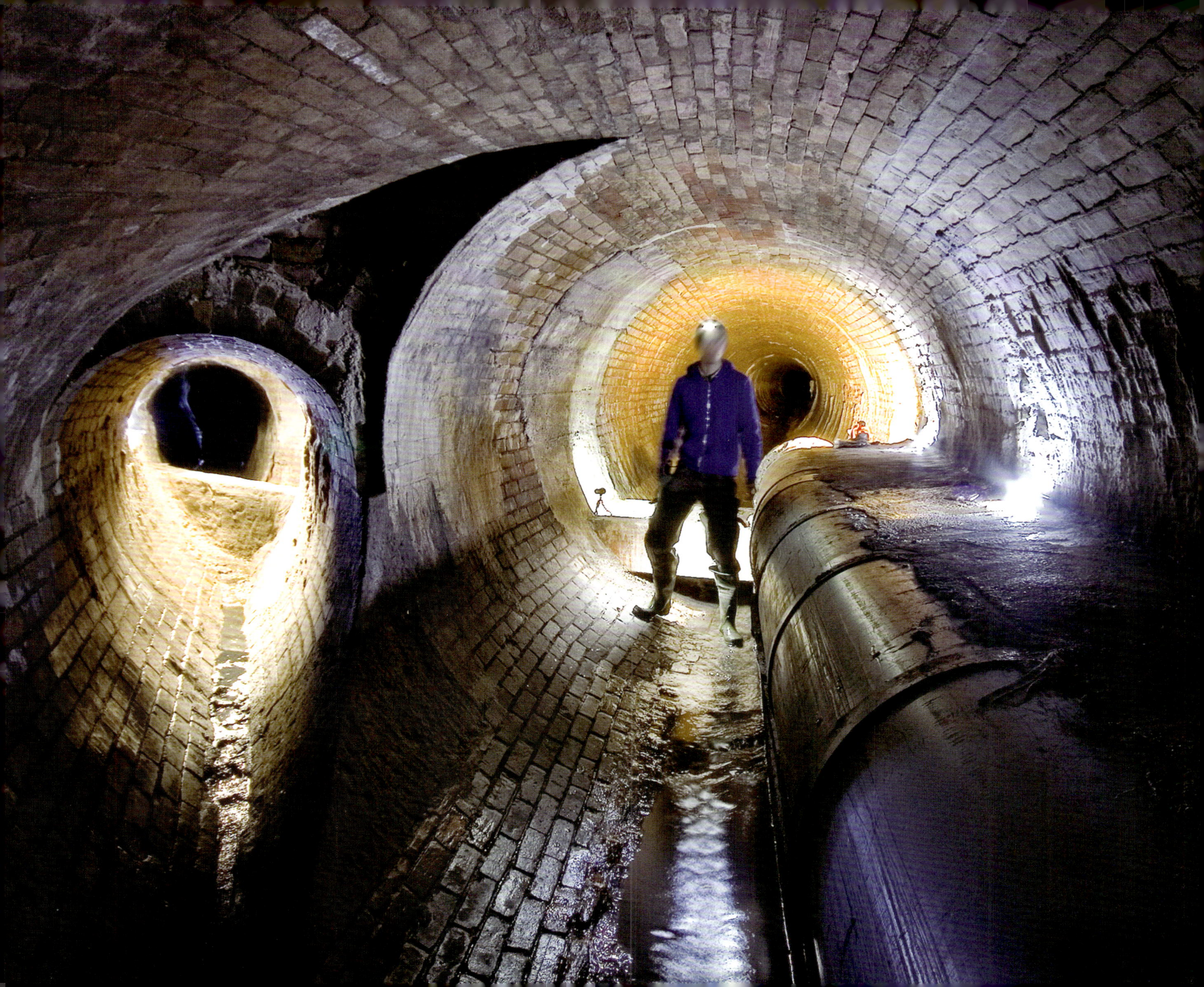

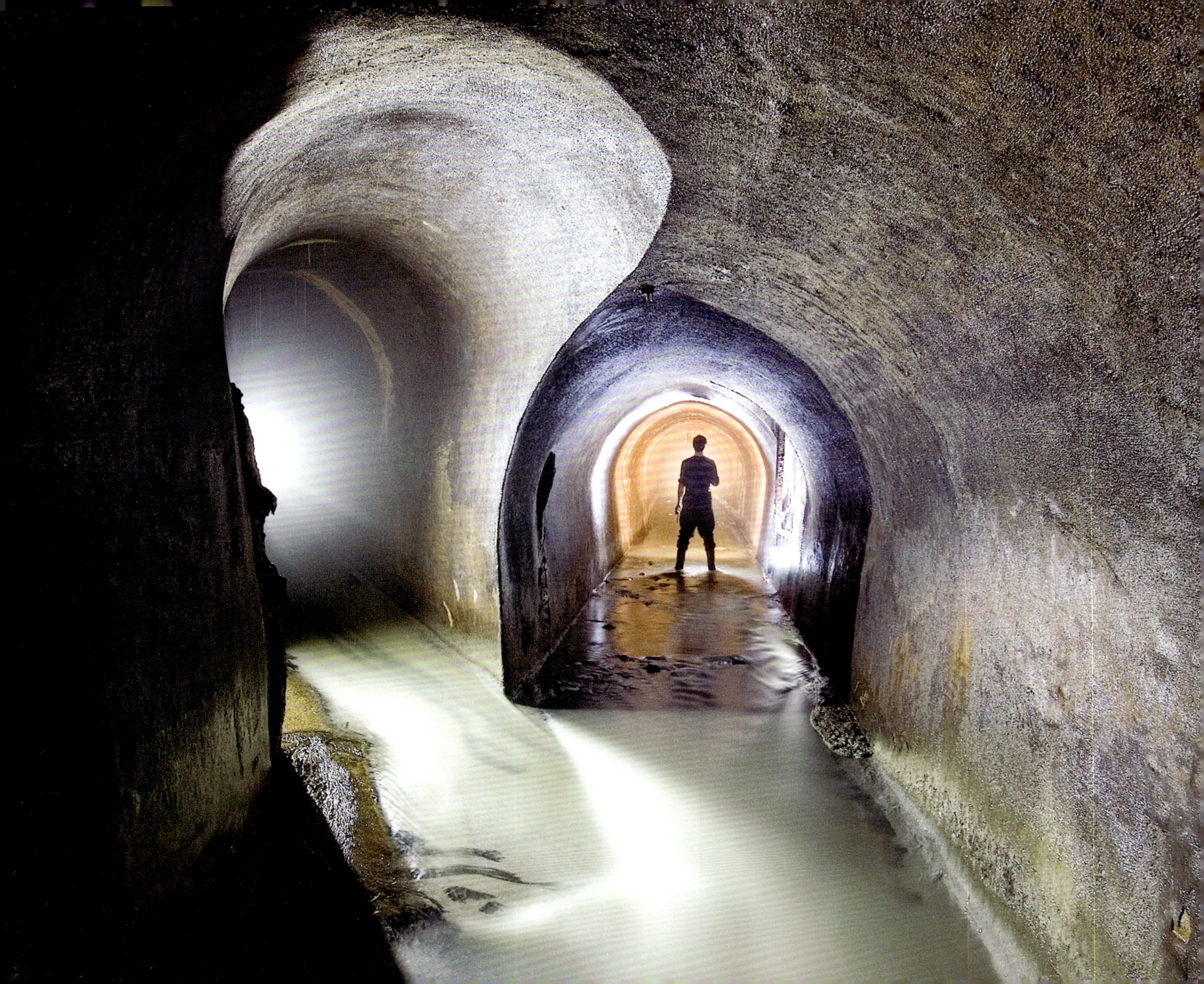

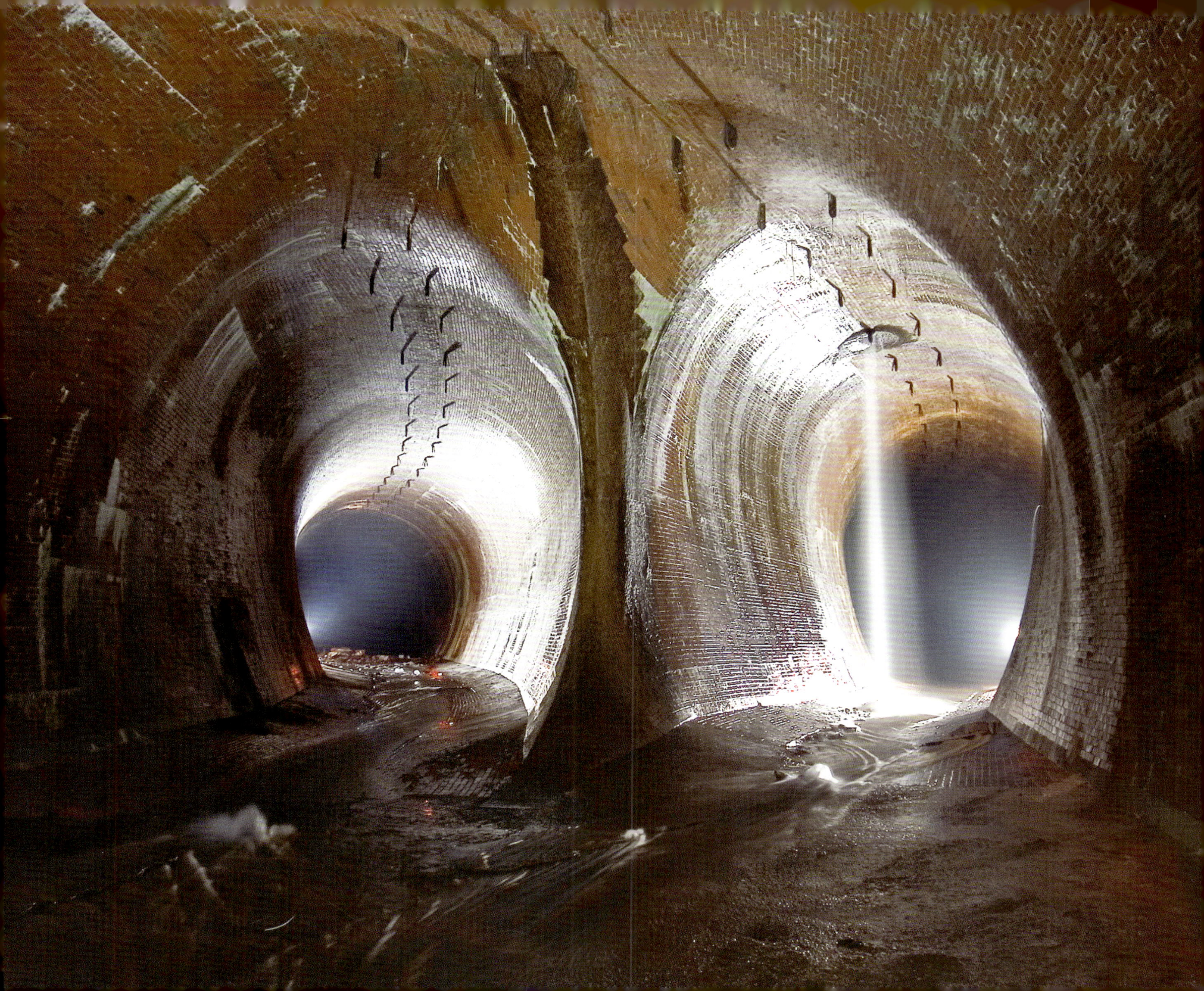

The elegant and the abrupt

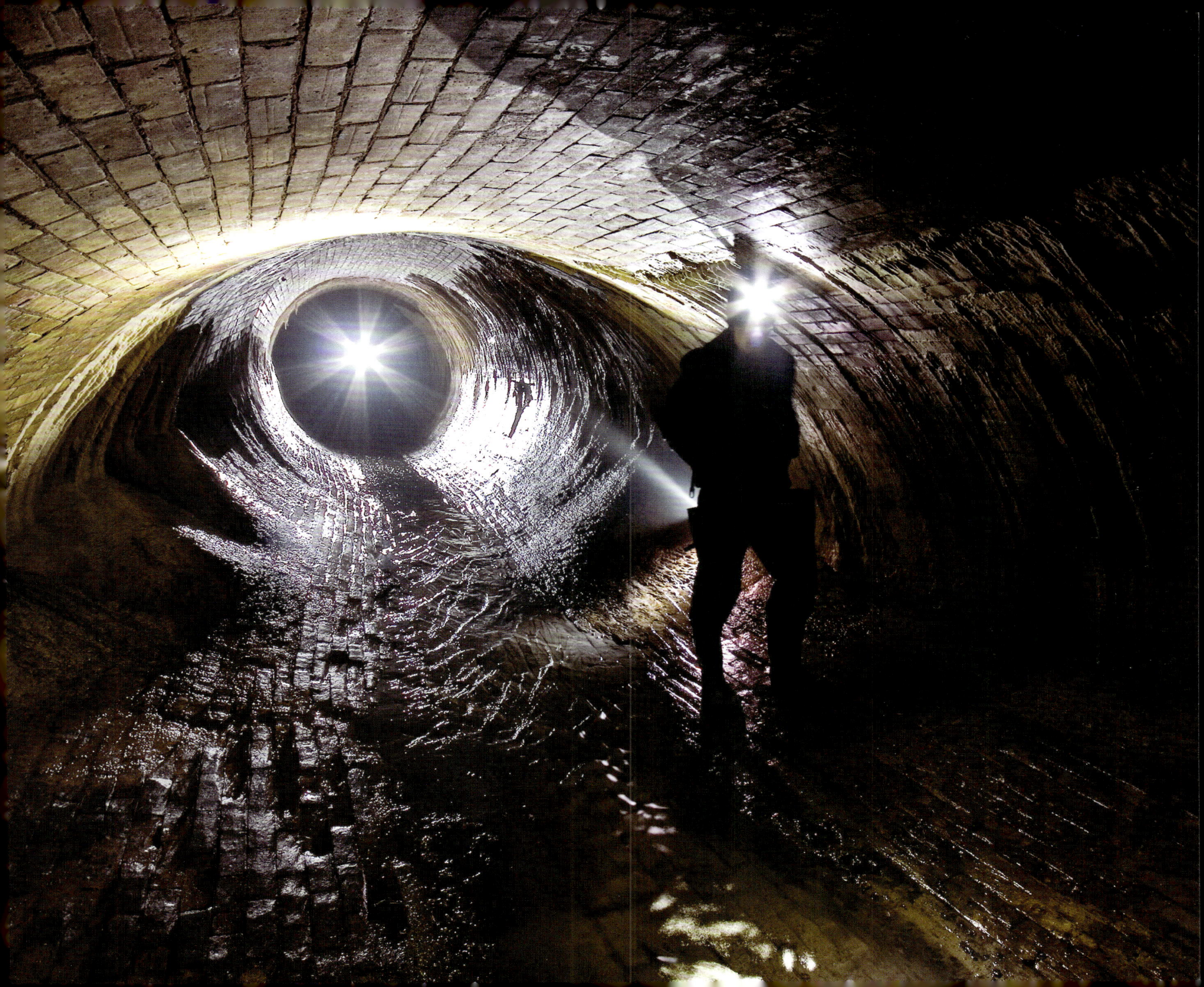

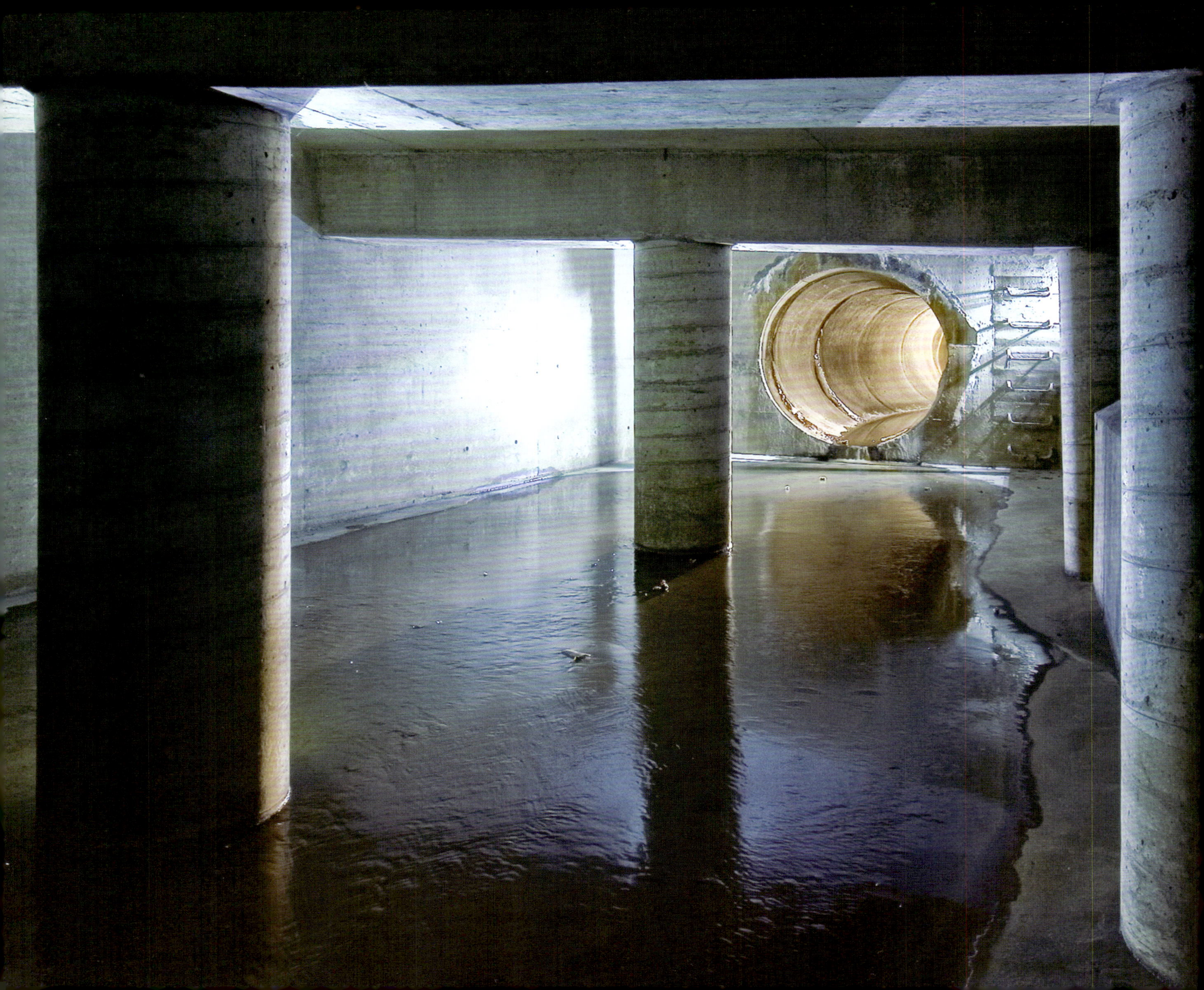

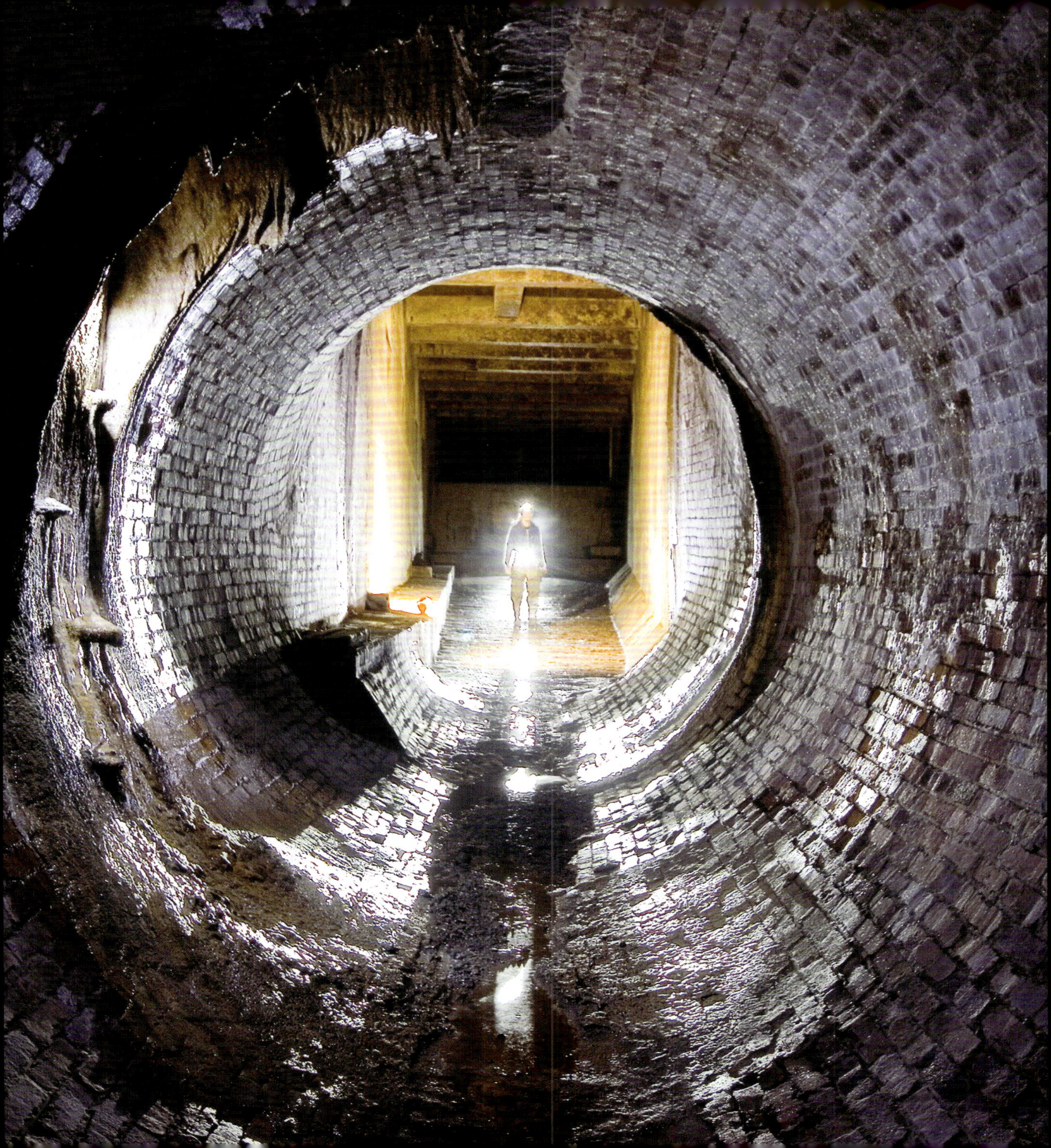

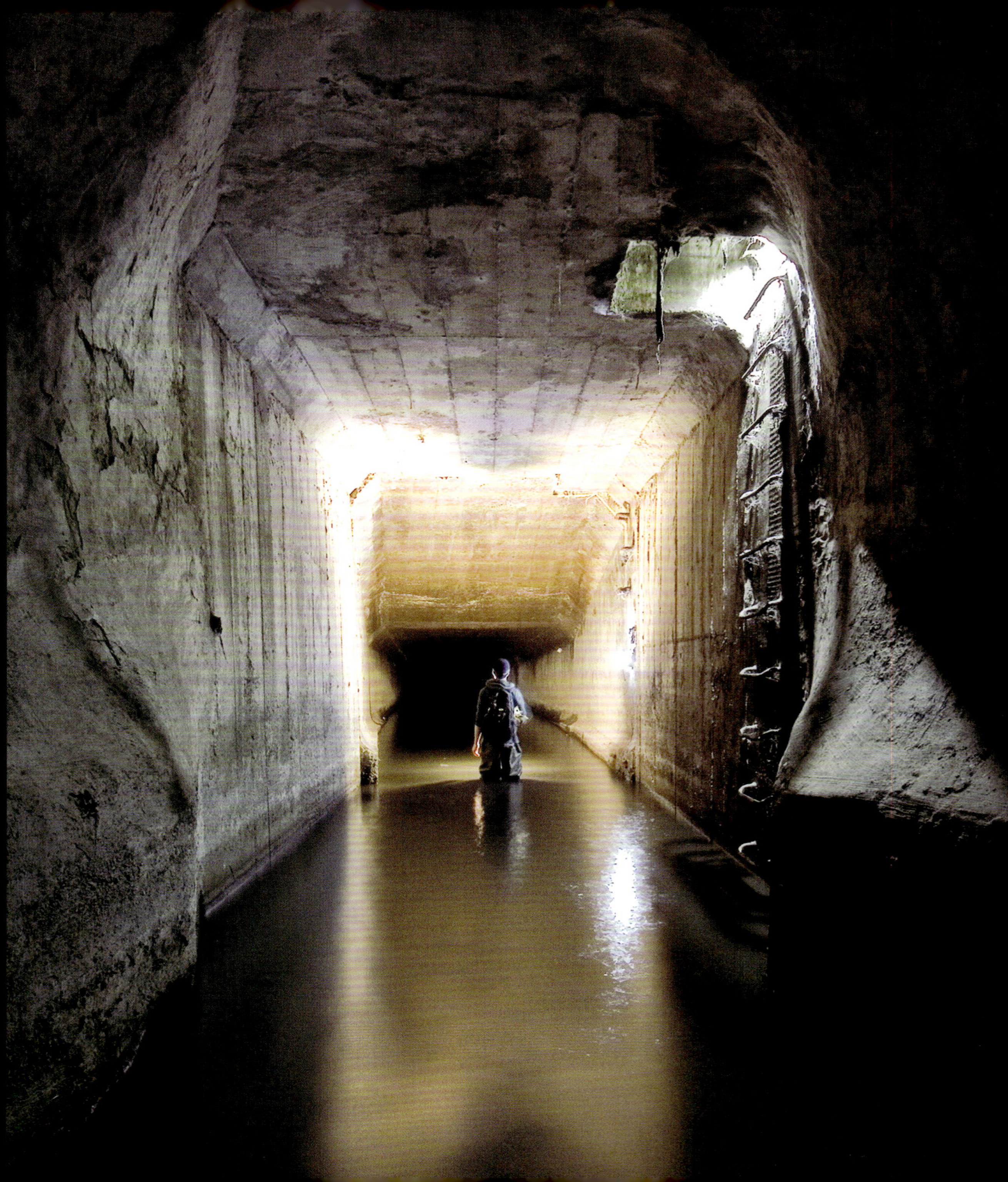

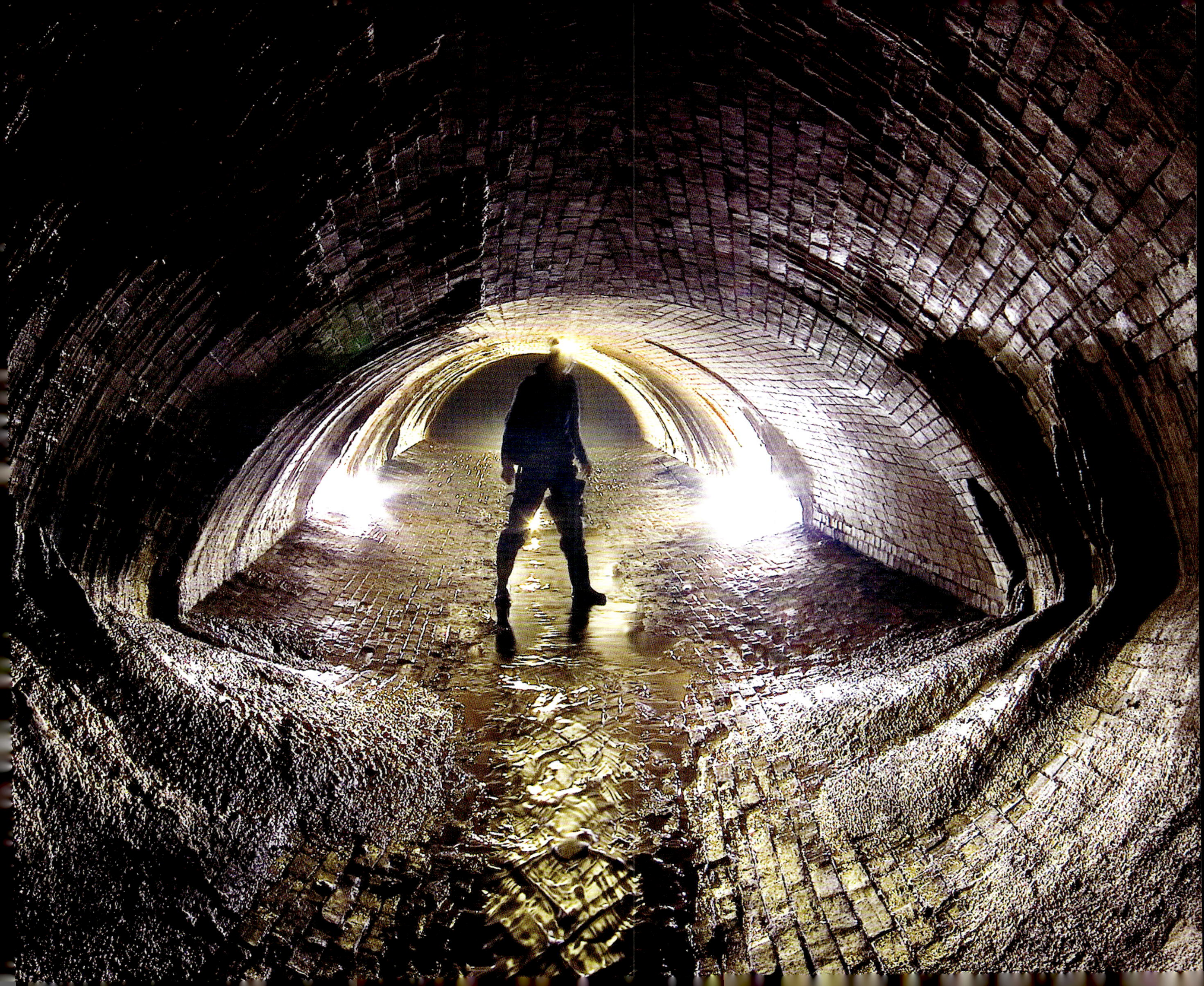

Elevations

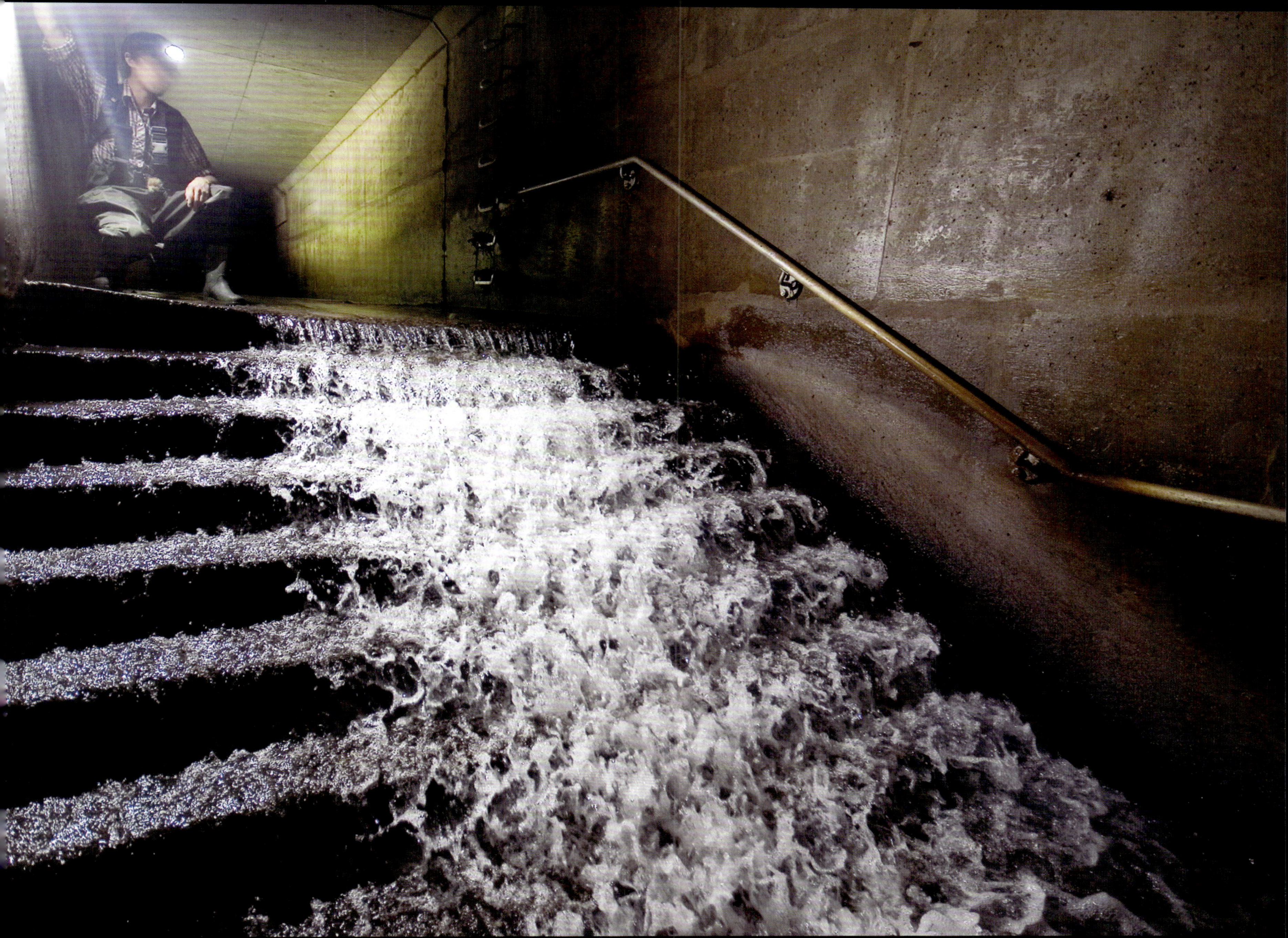

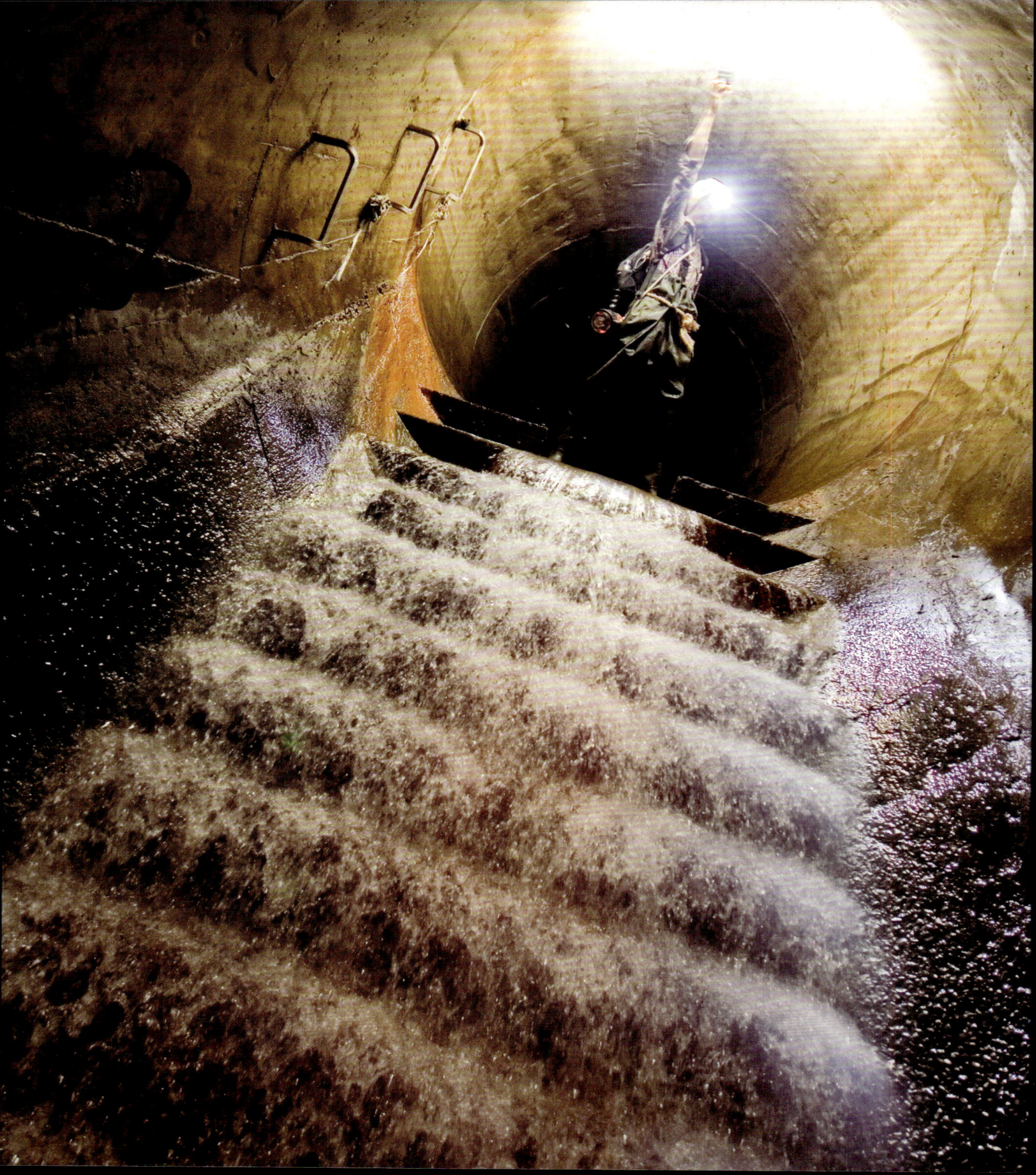

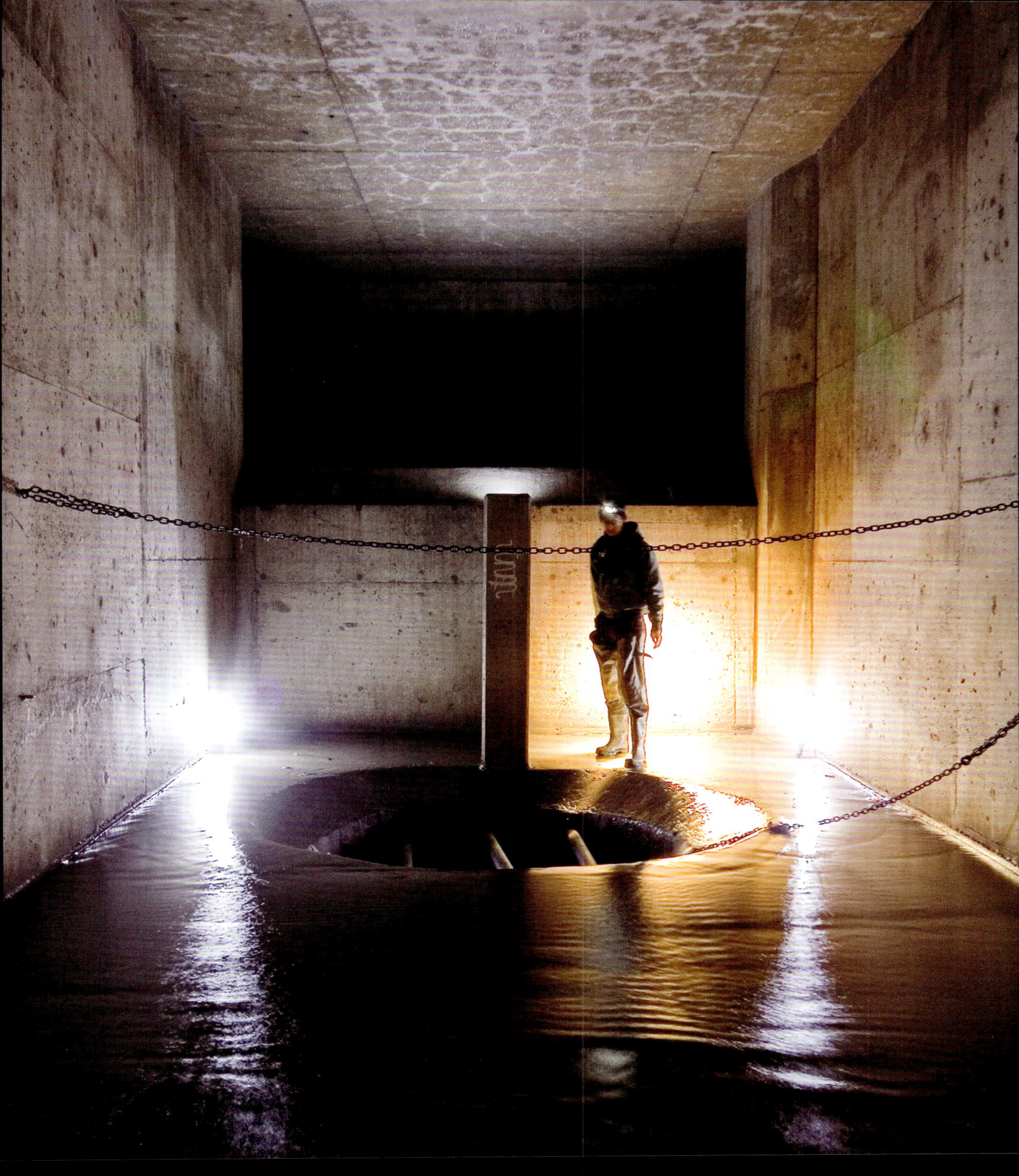

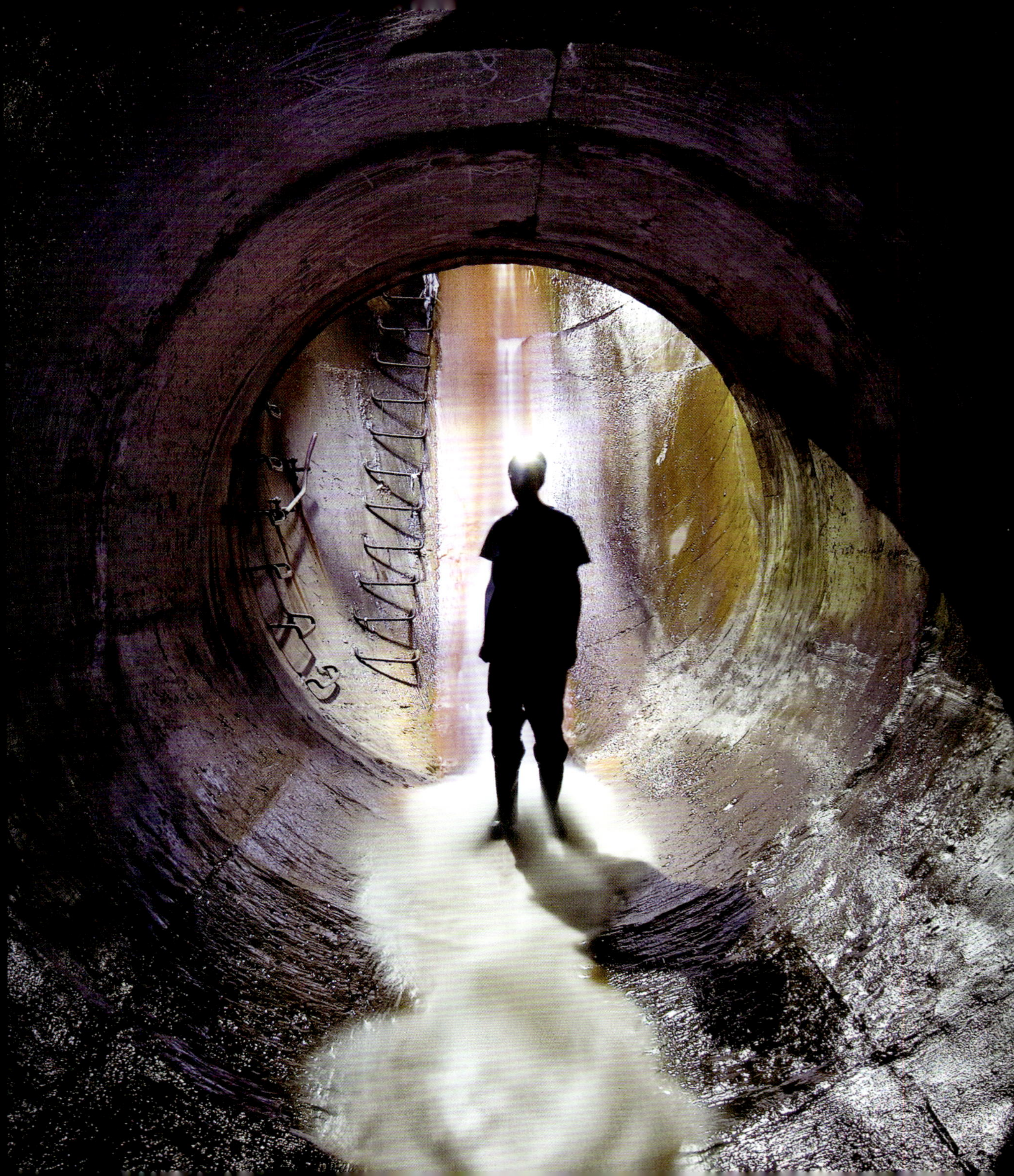

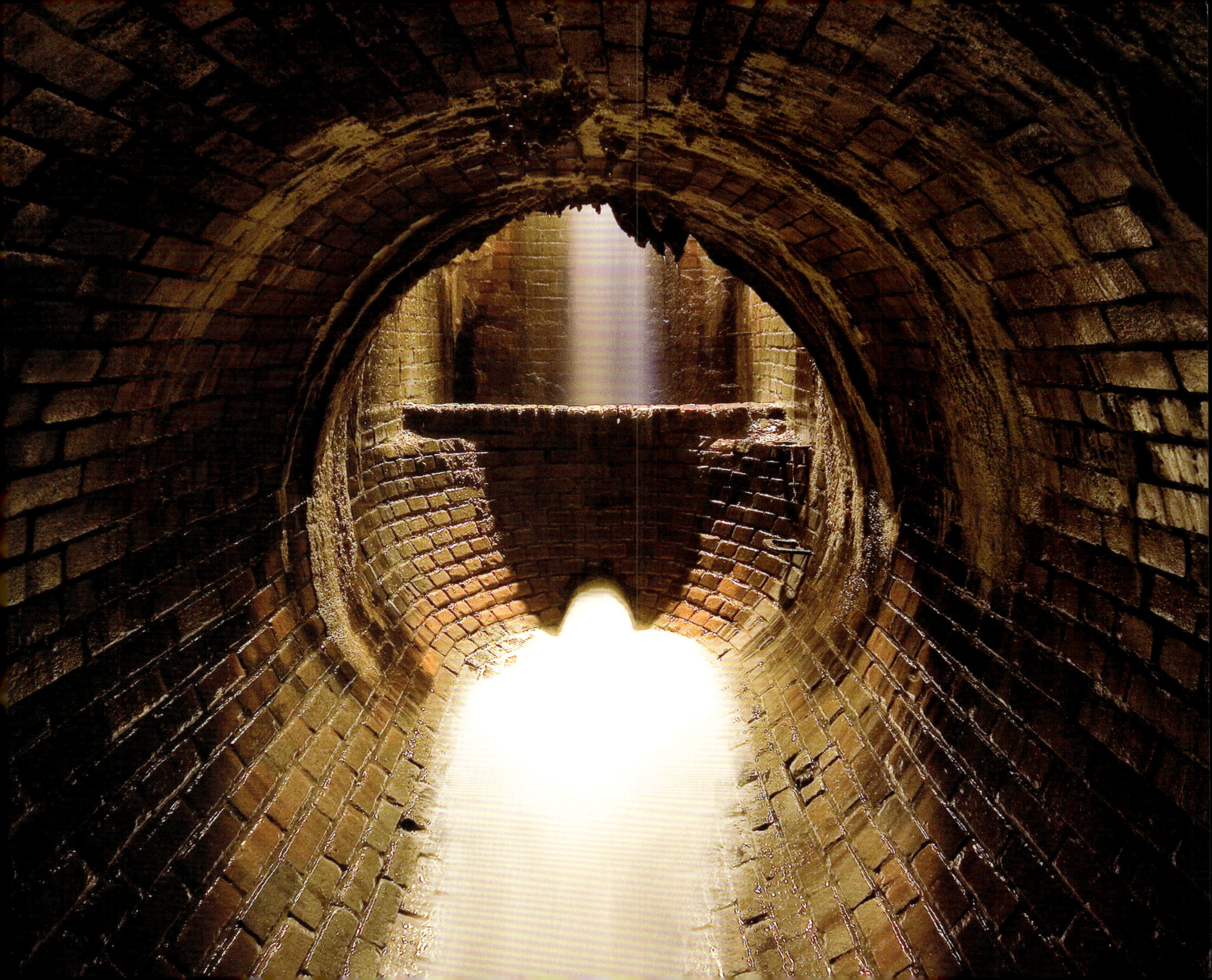

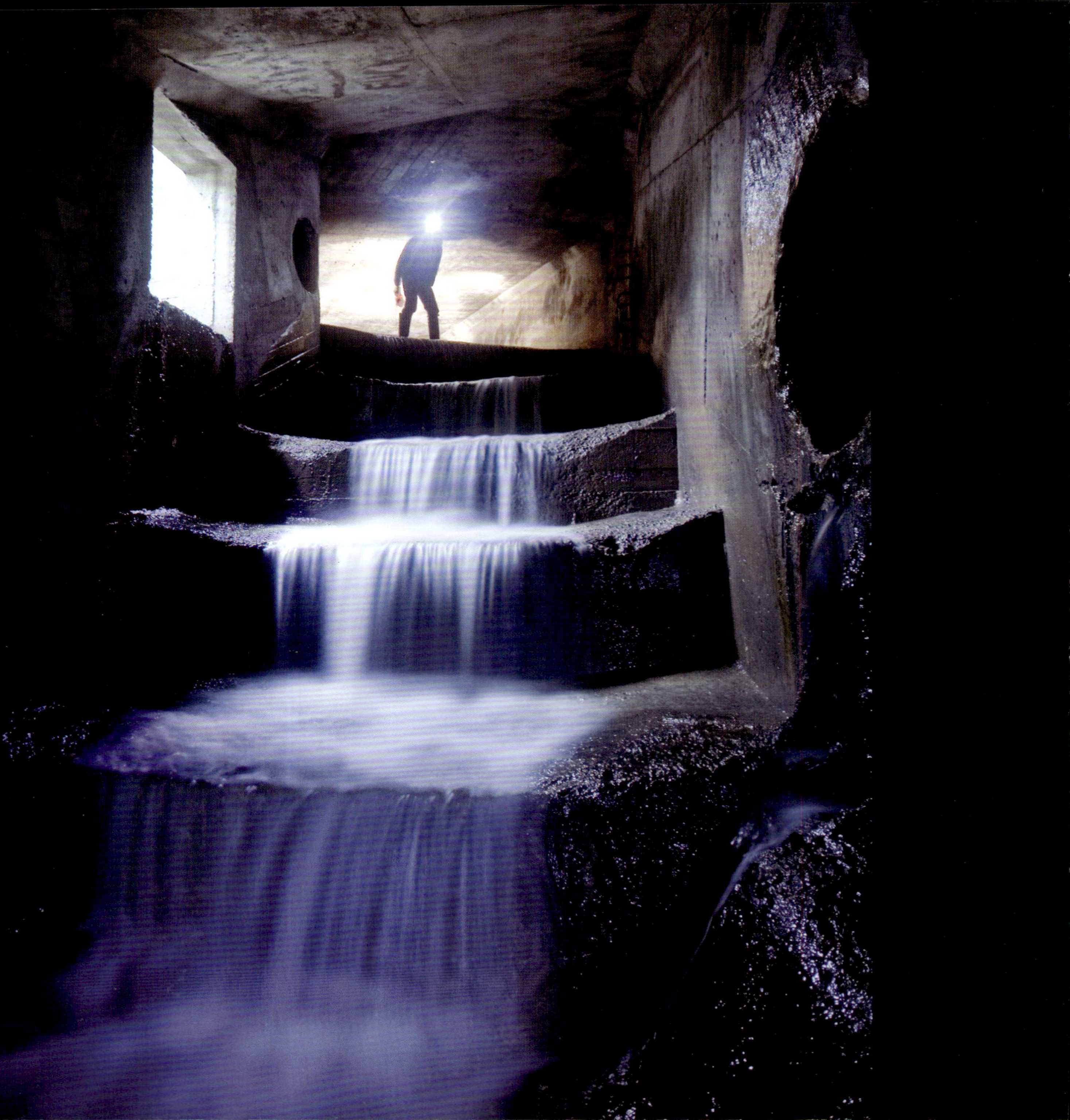

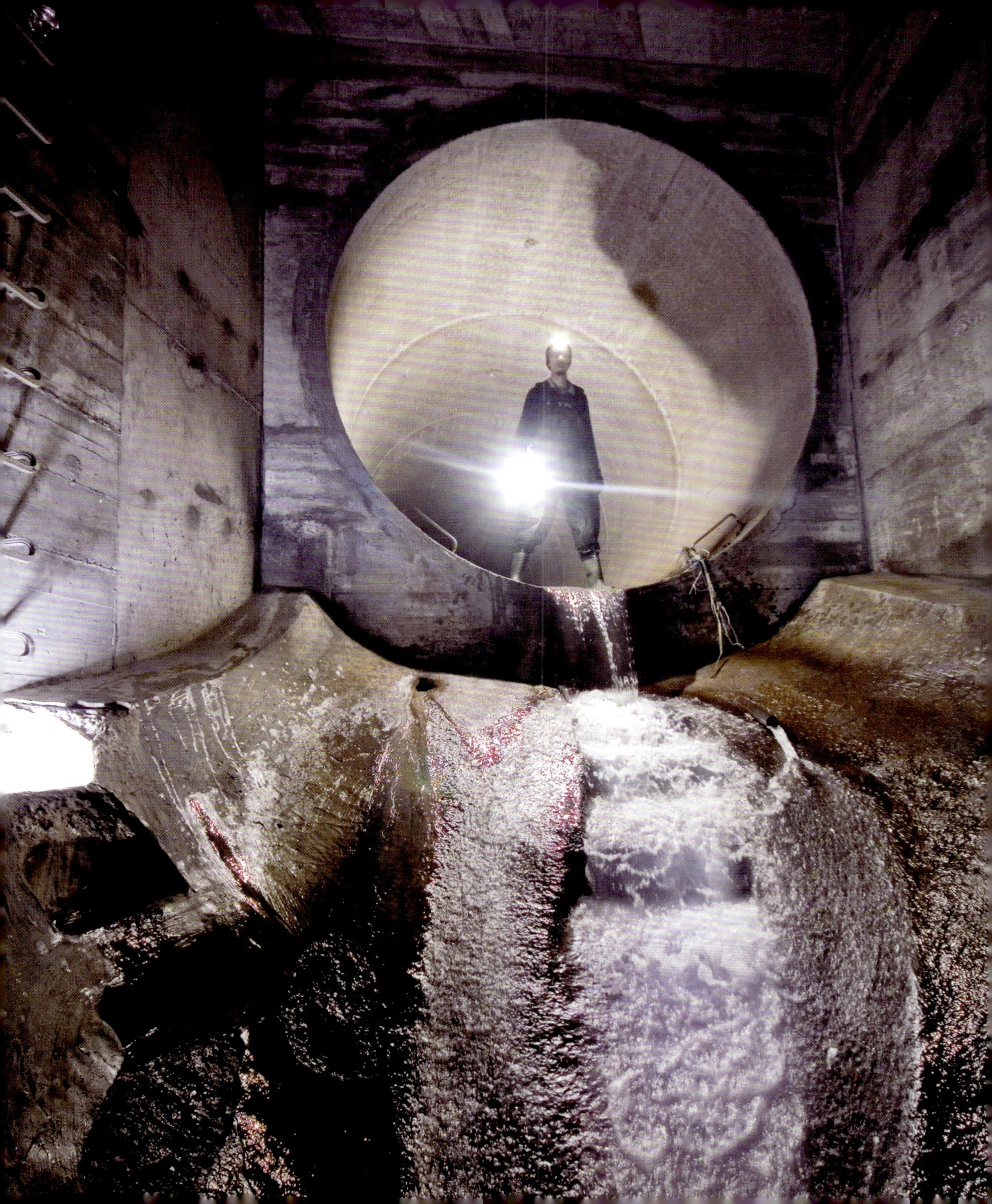

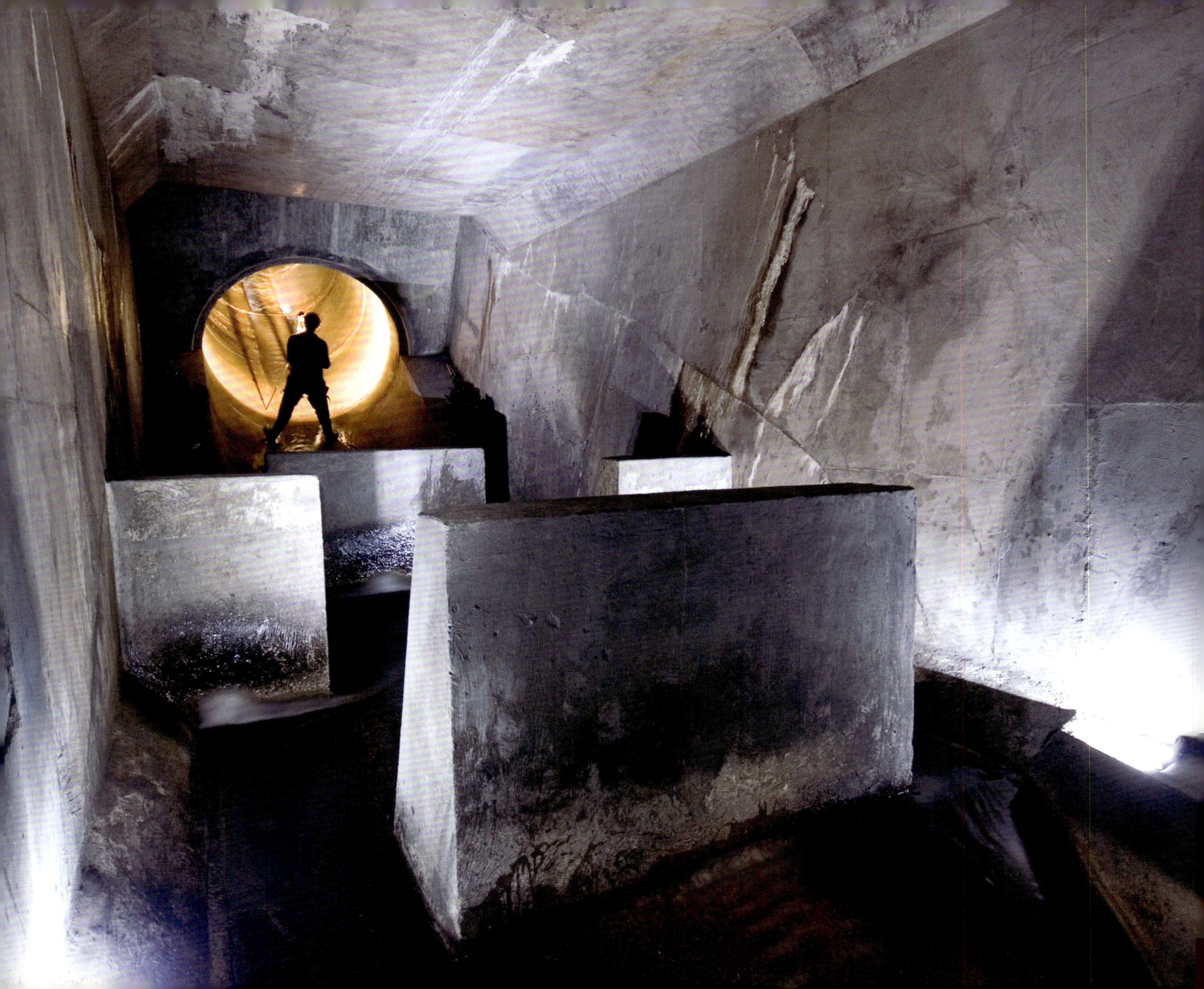

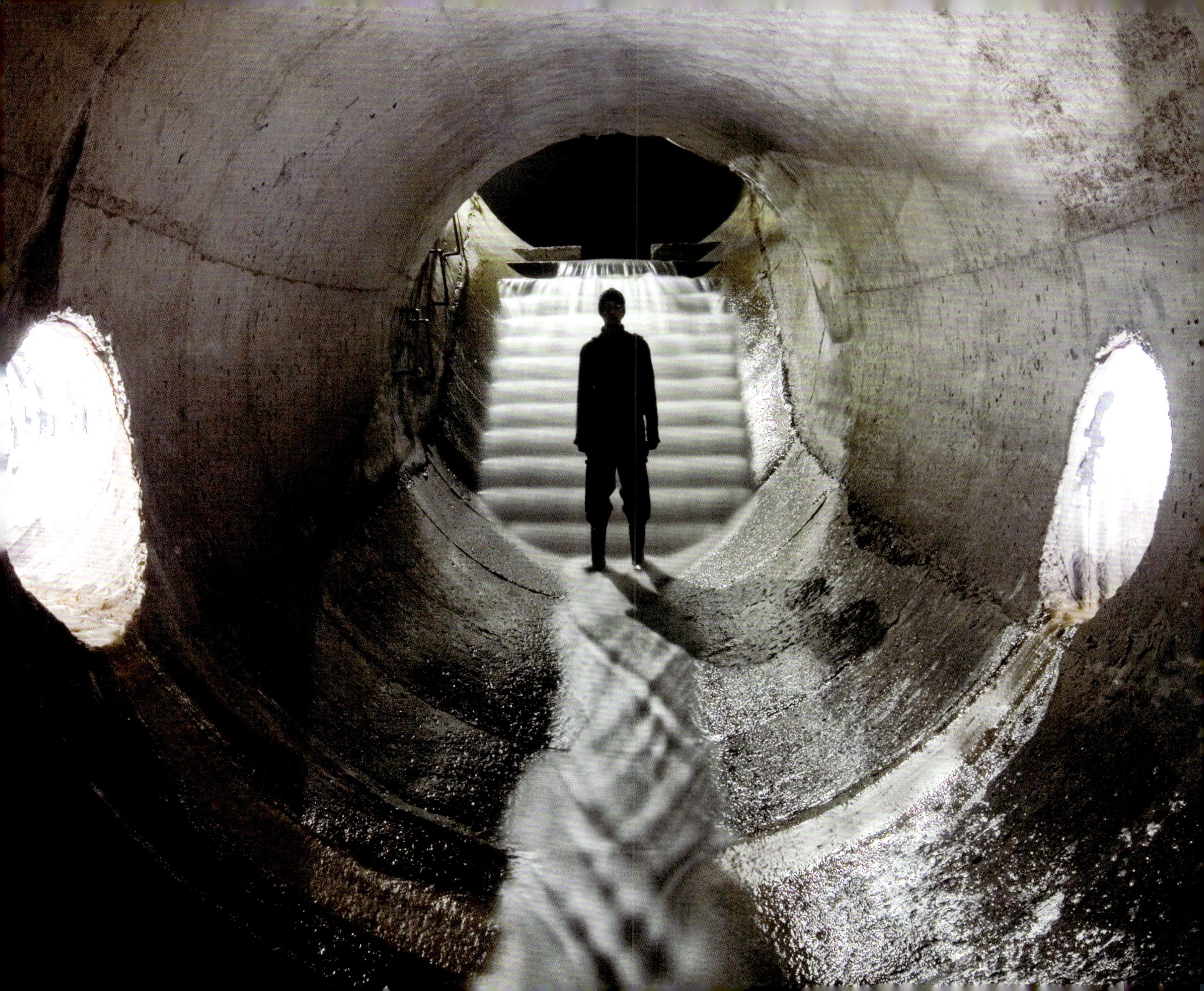

Generations

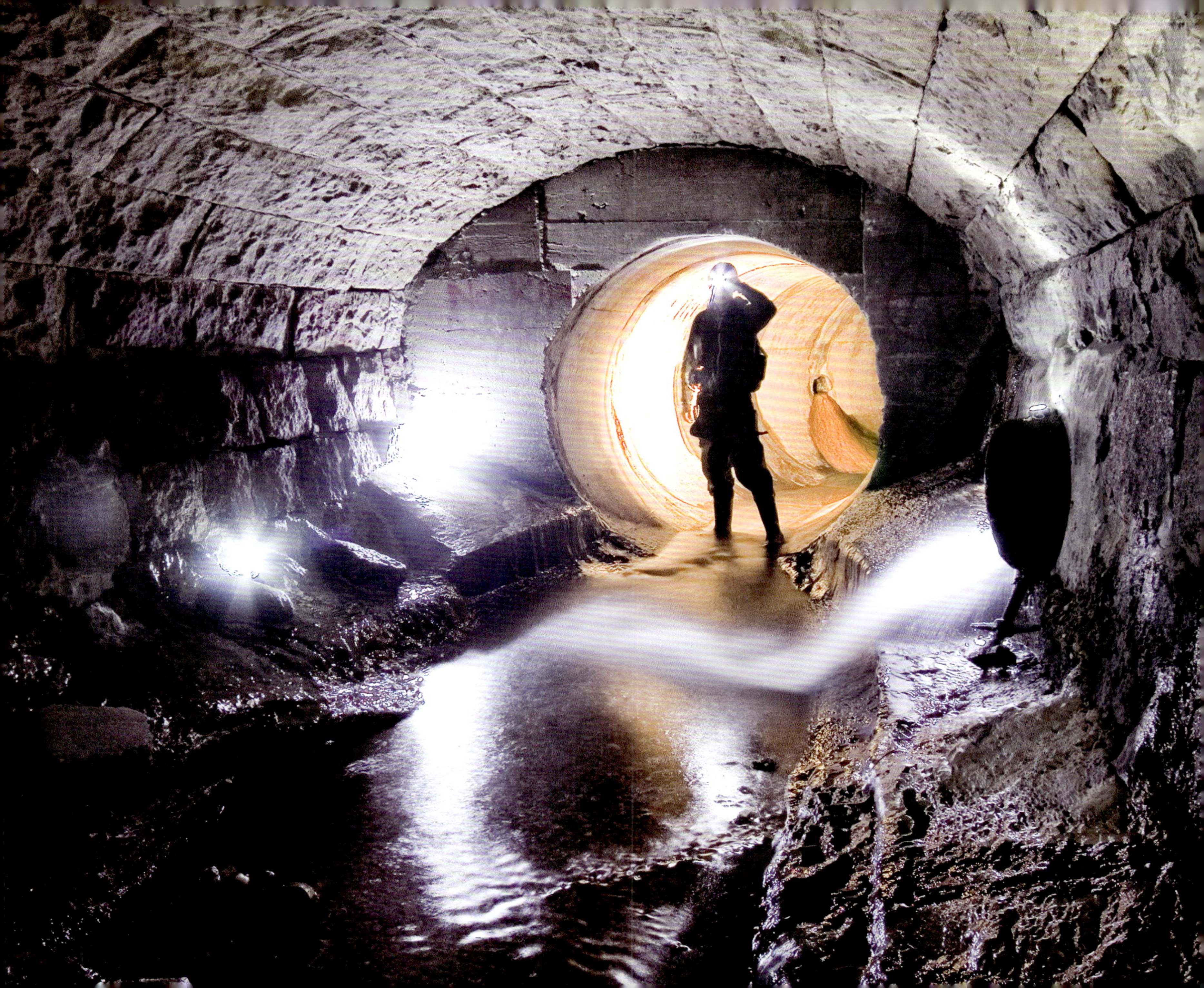

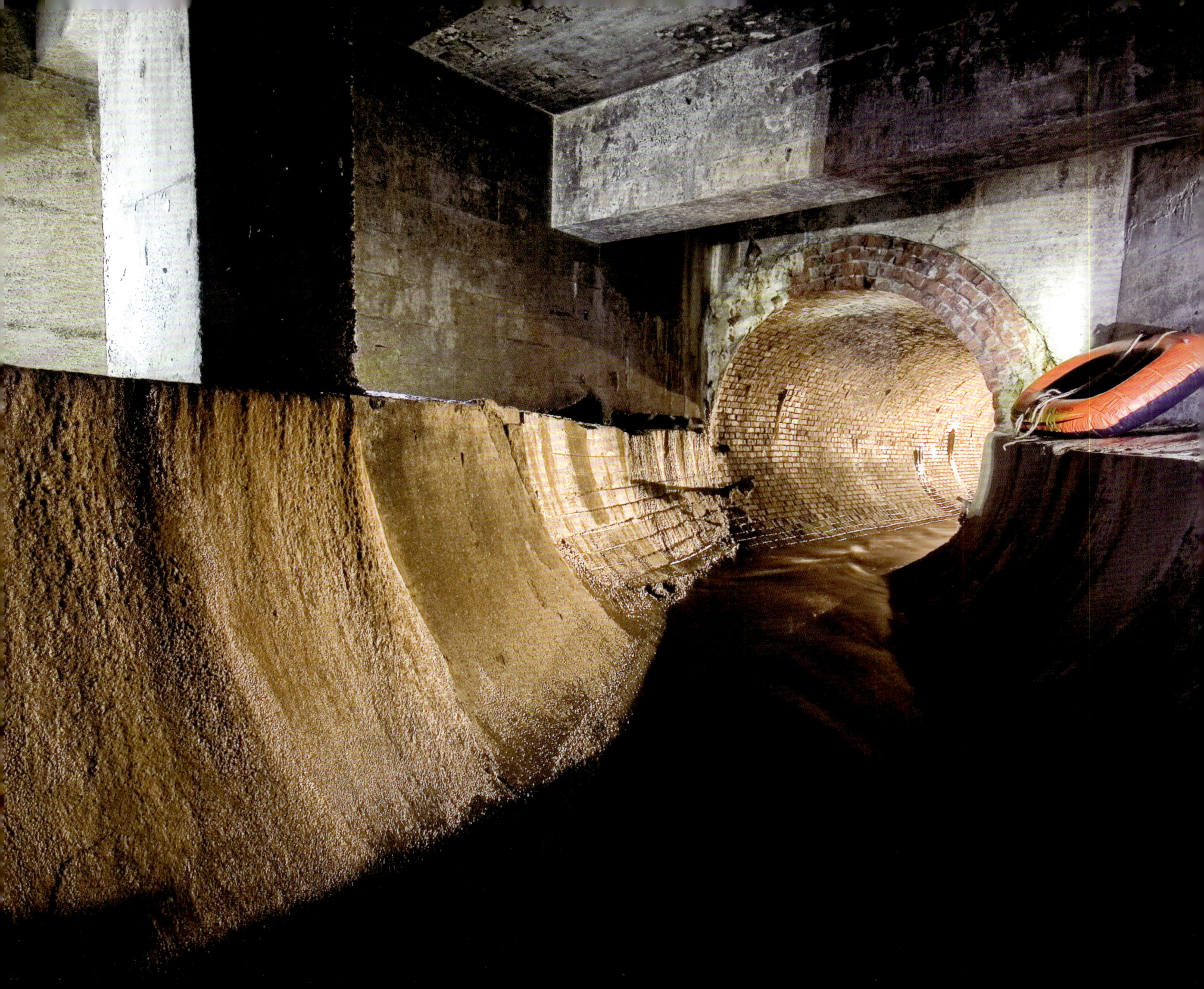

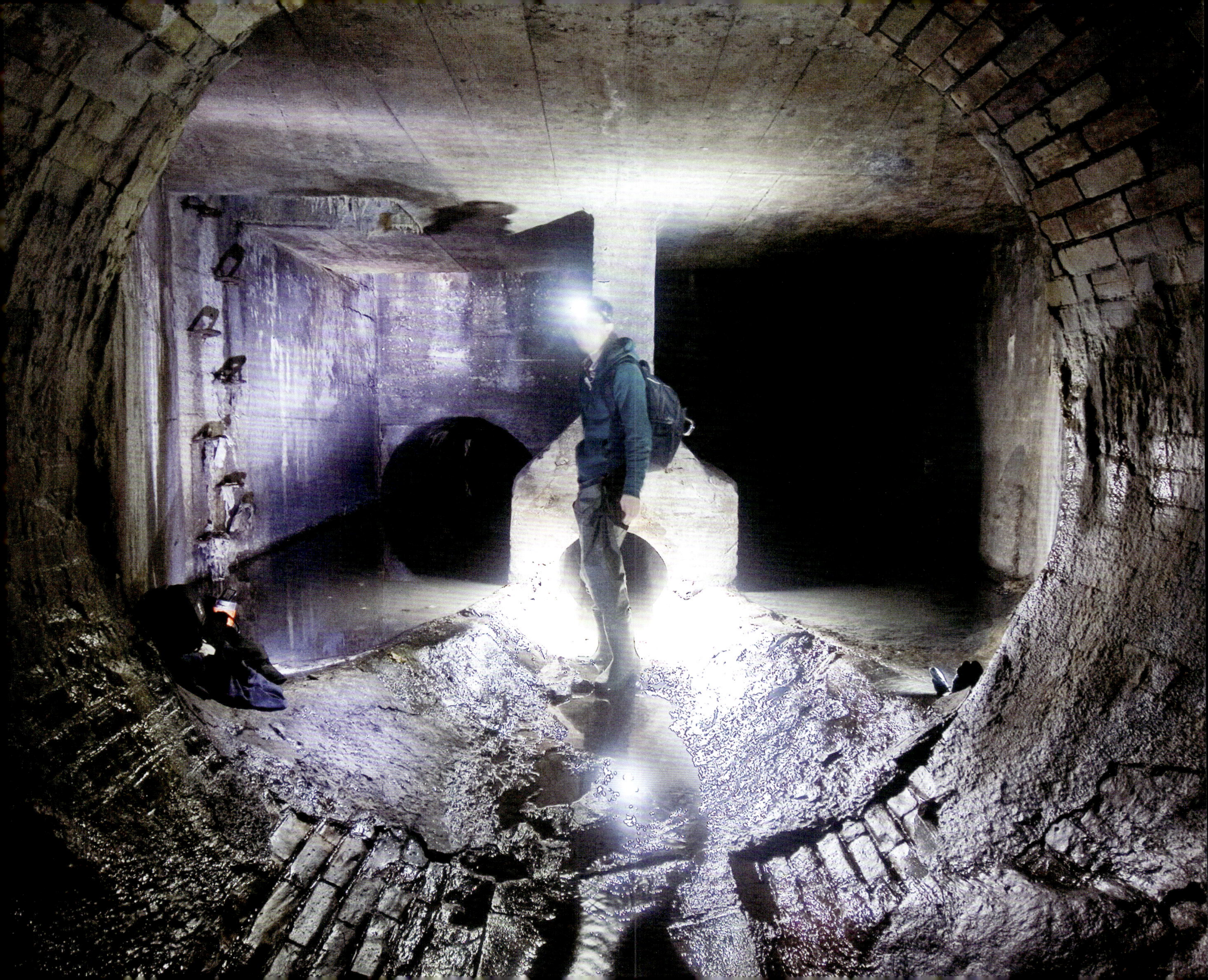

Overflow

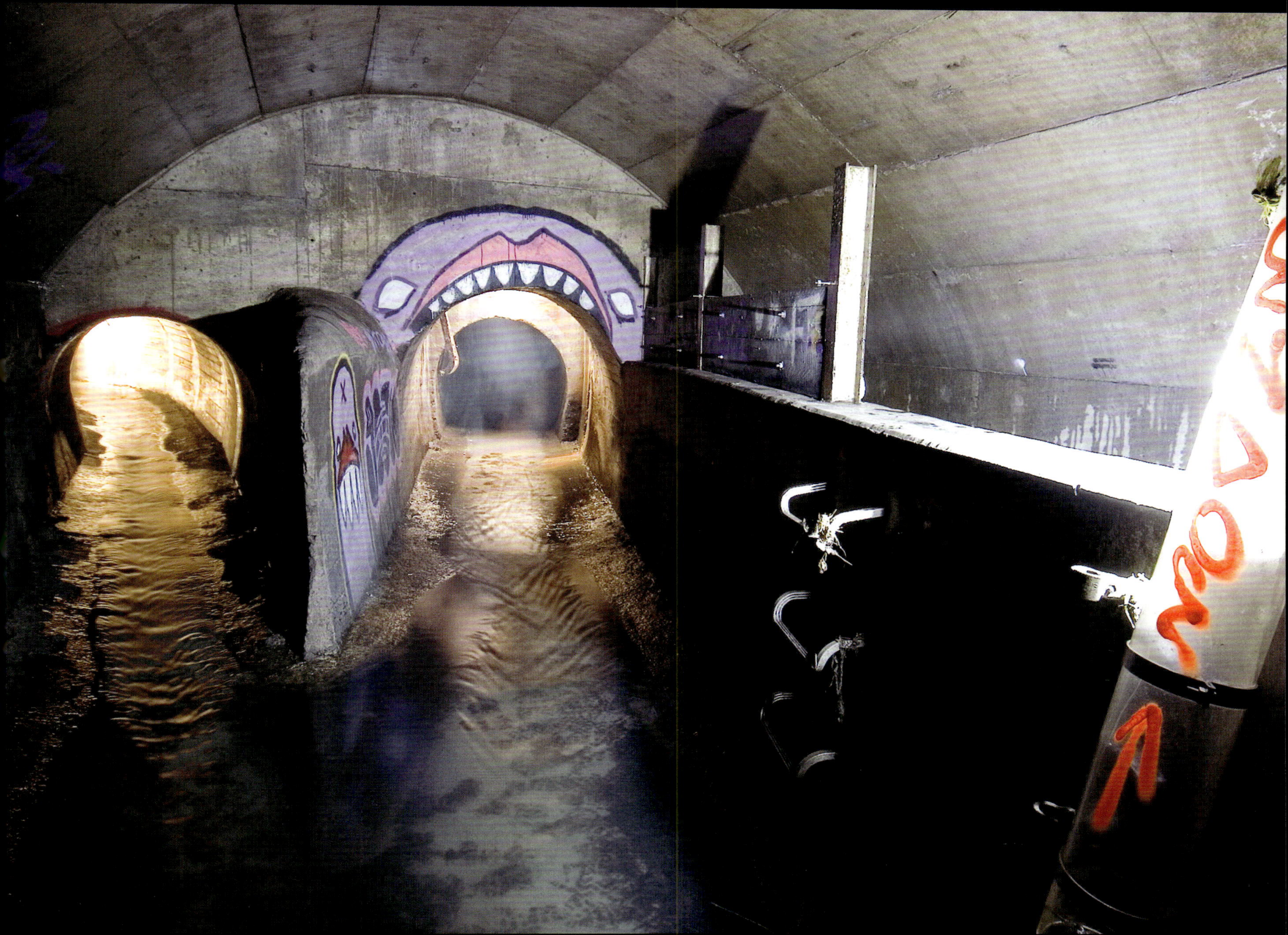

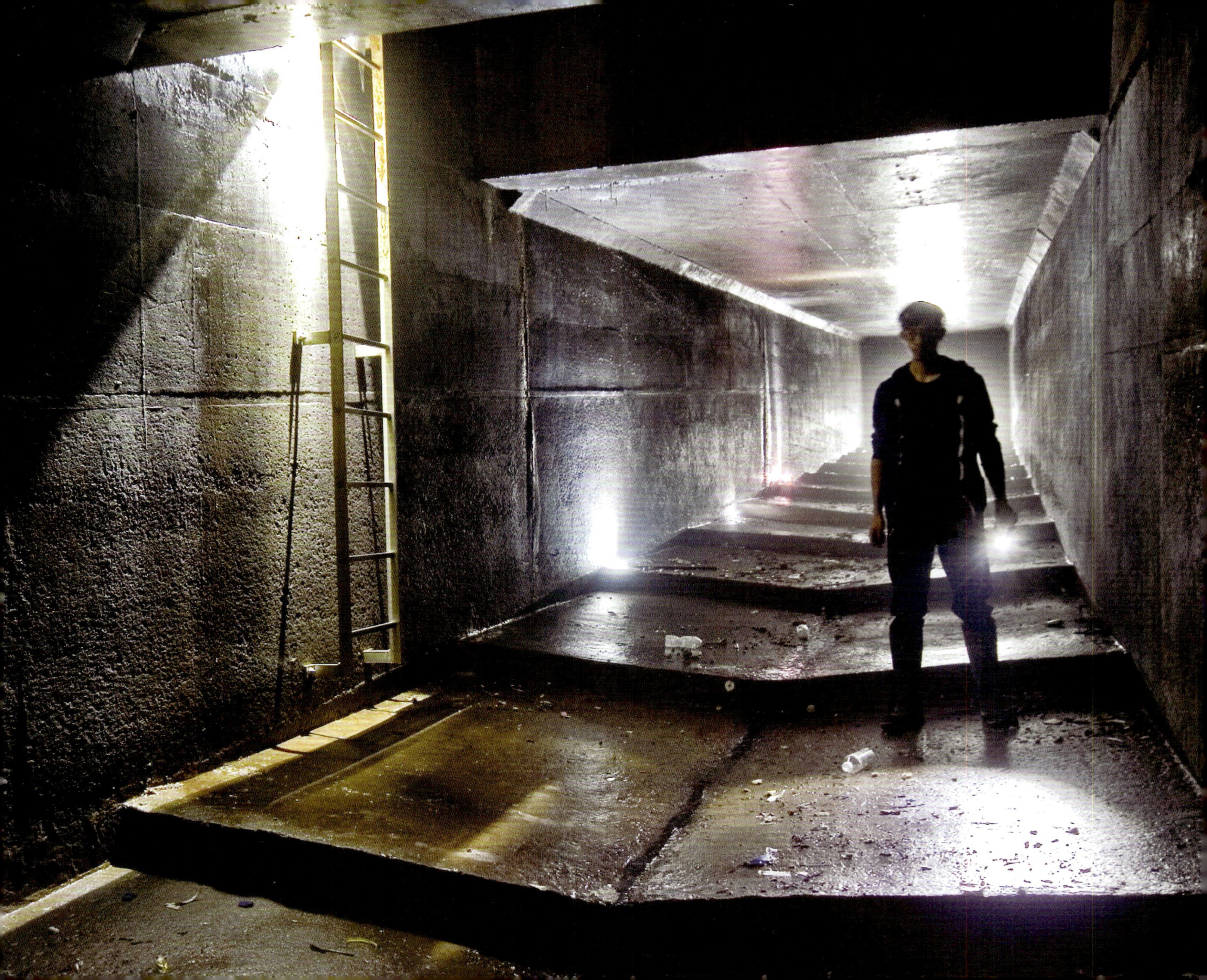

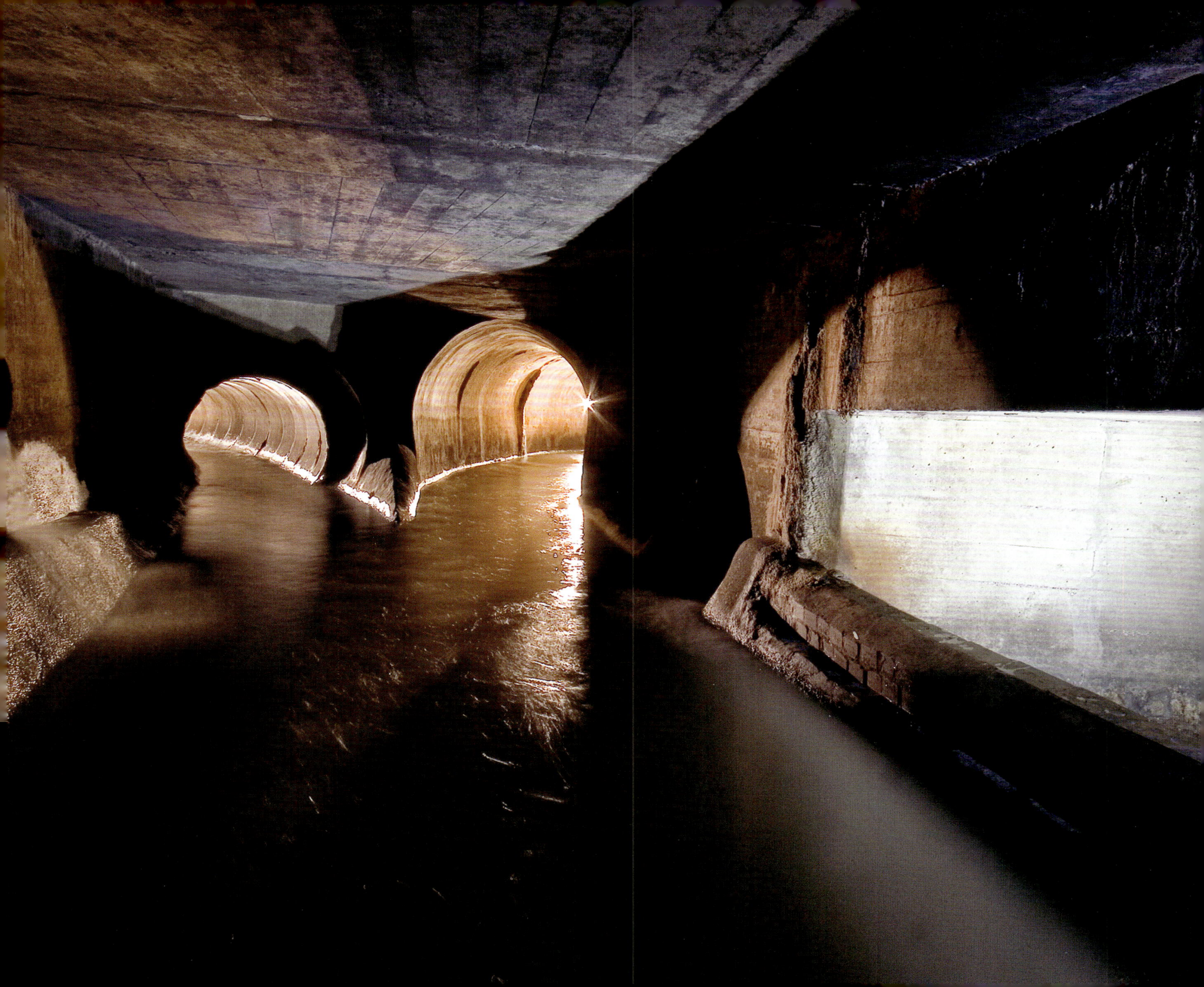

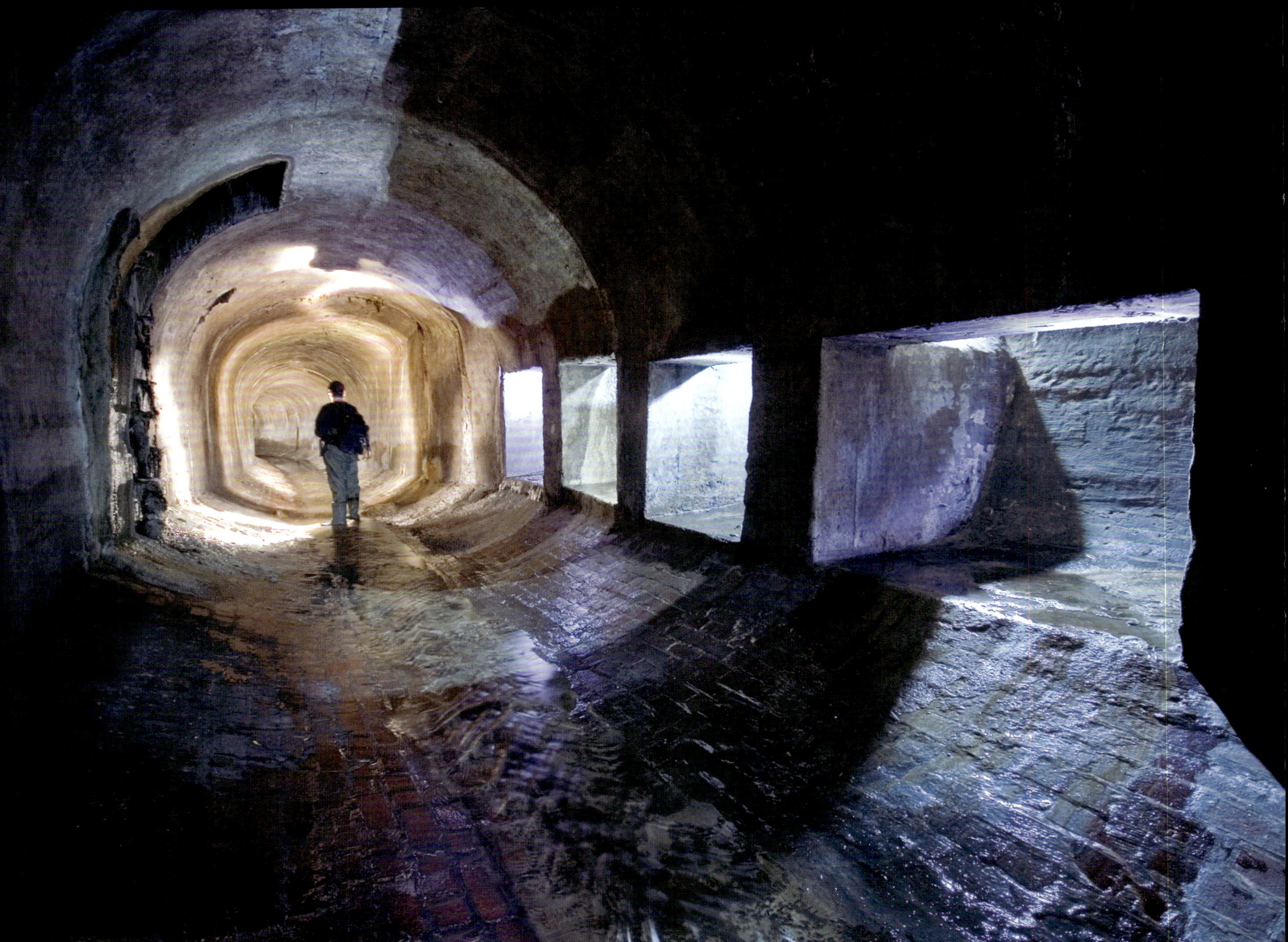

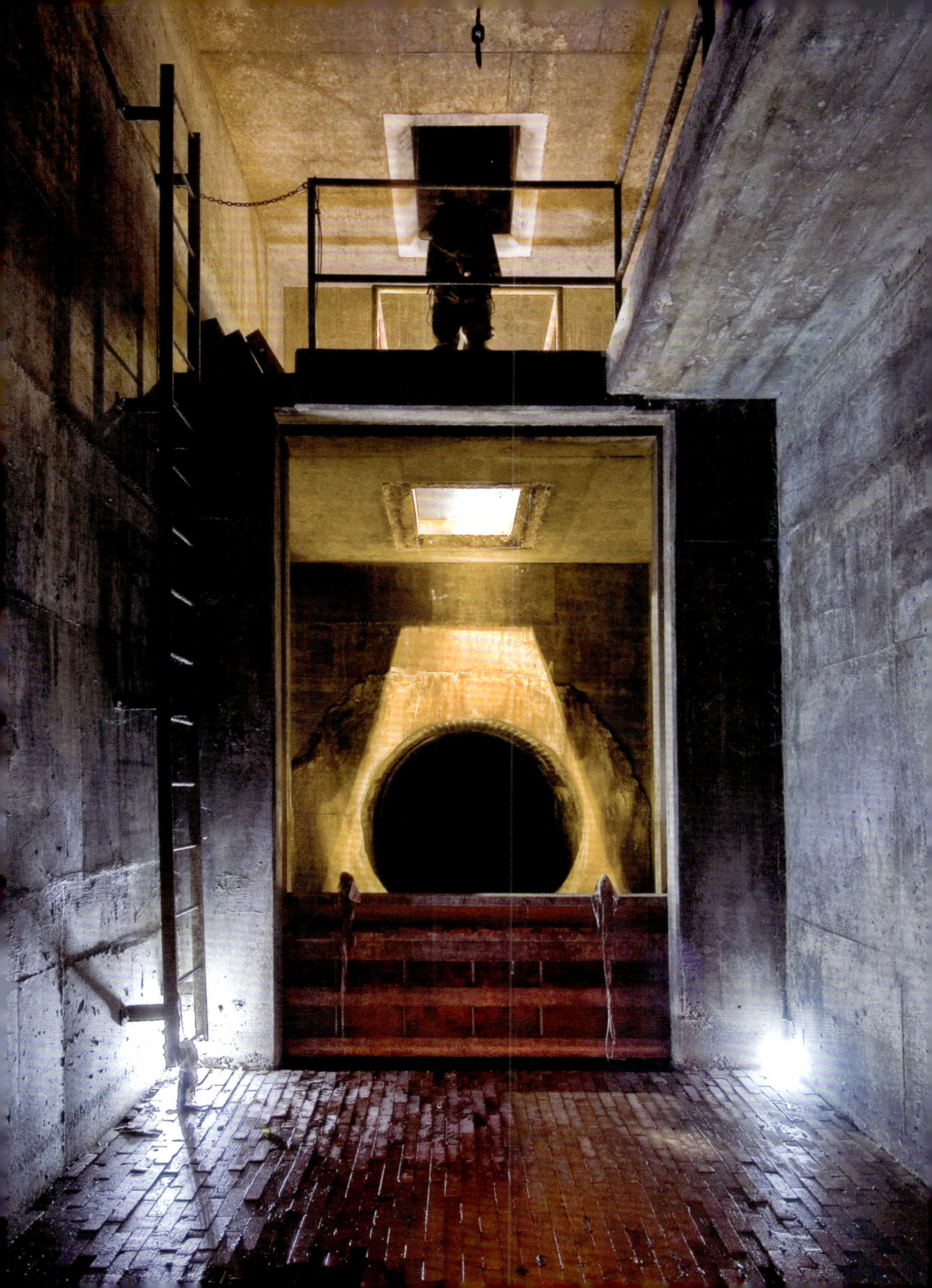

Access from above

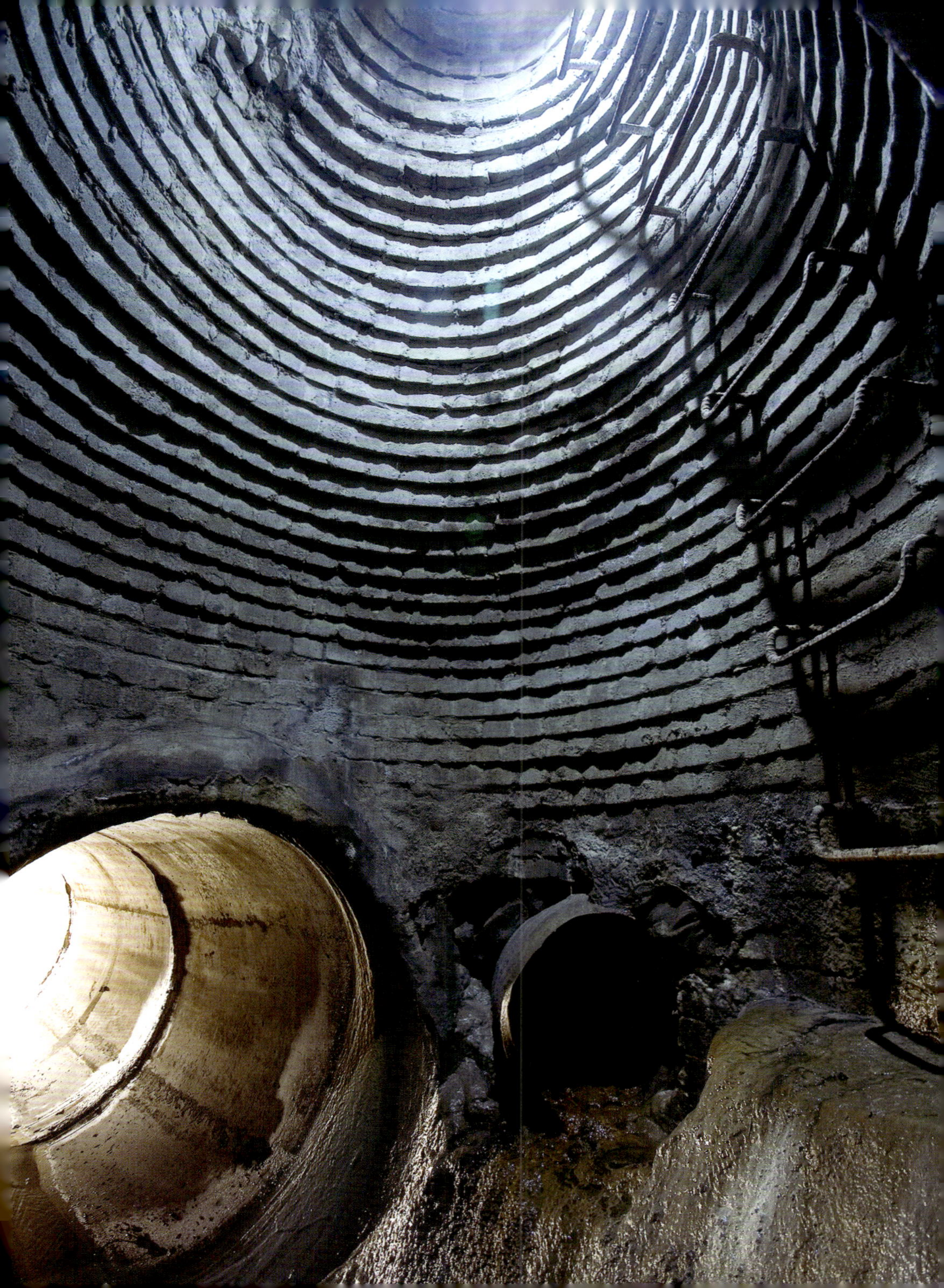

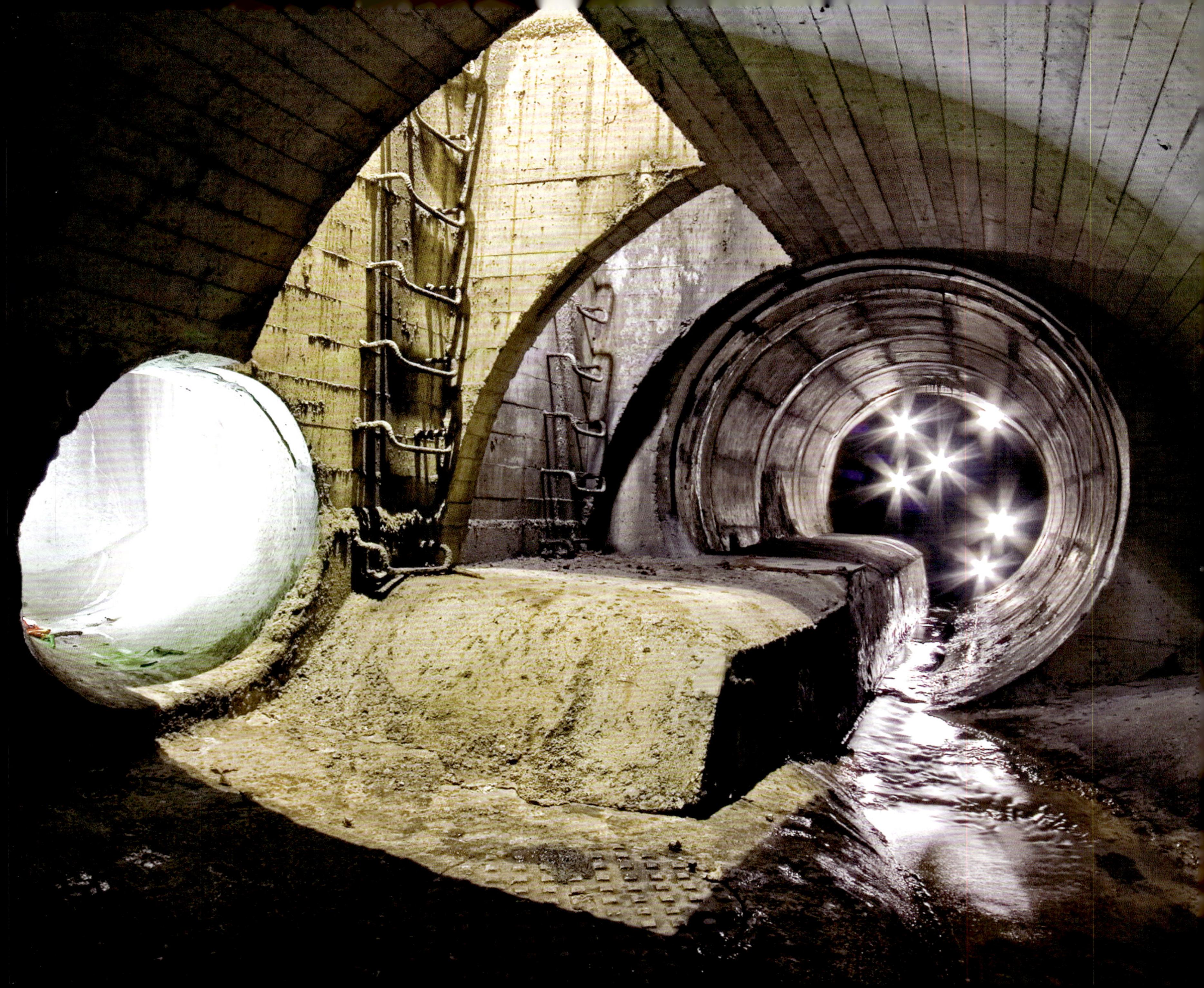

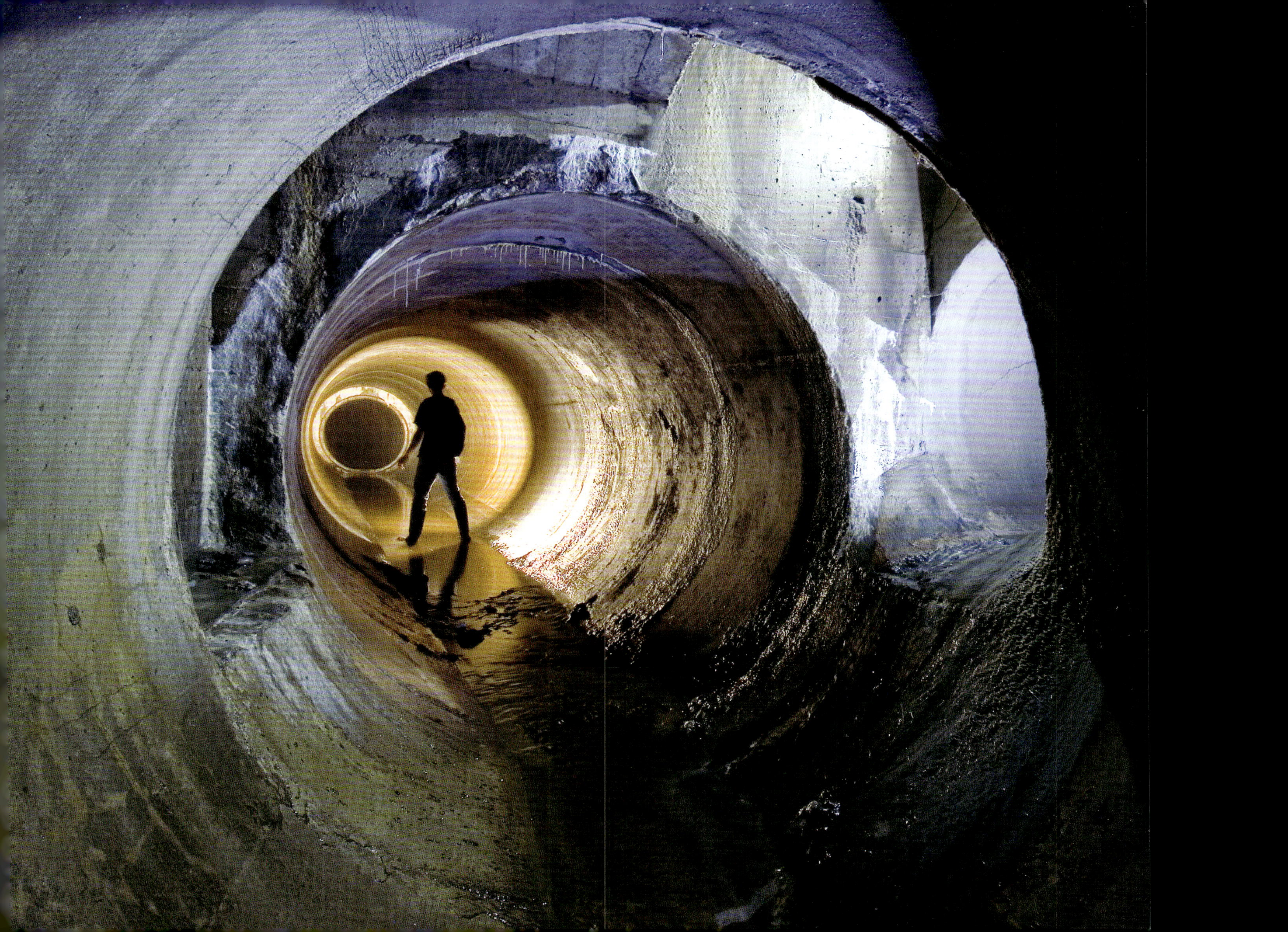

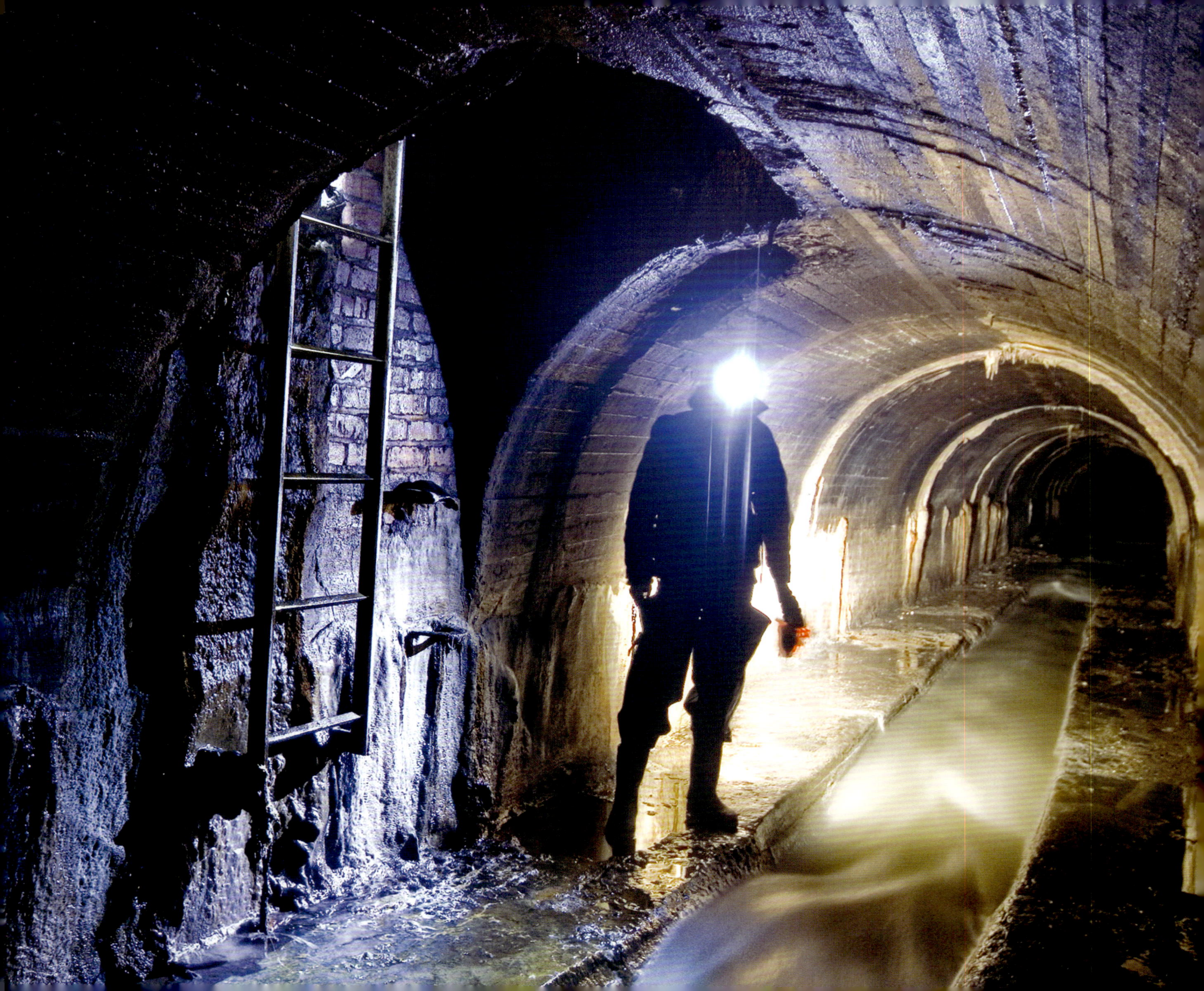

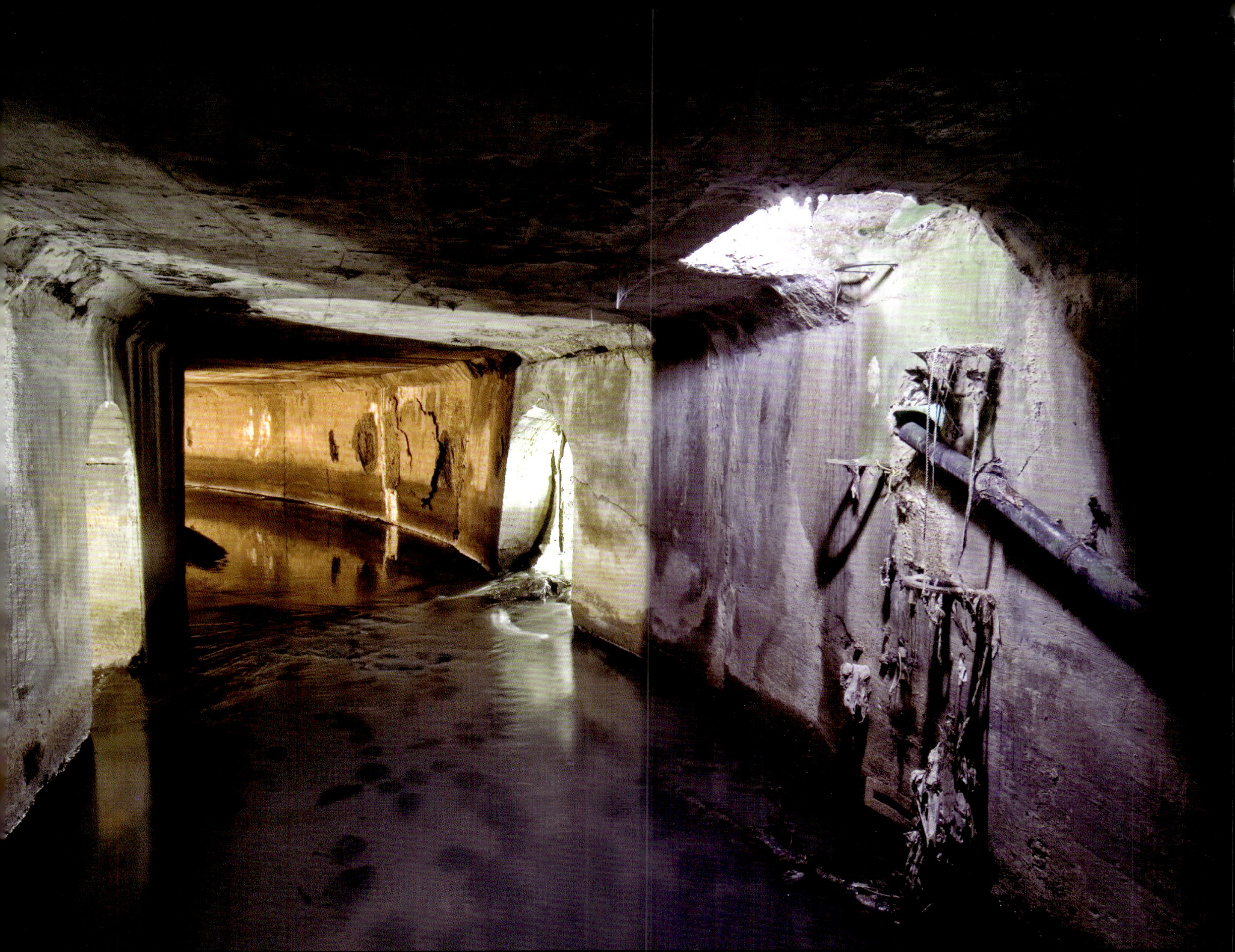

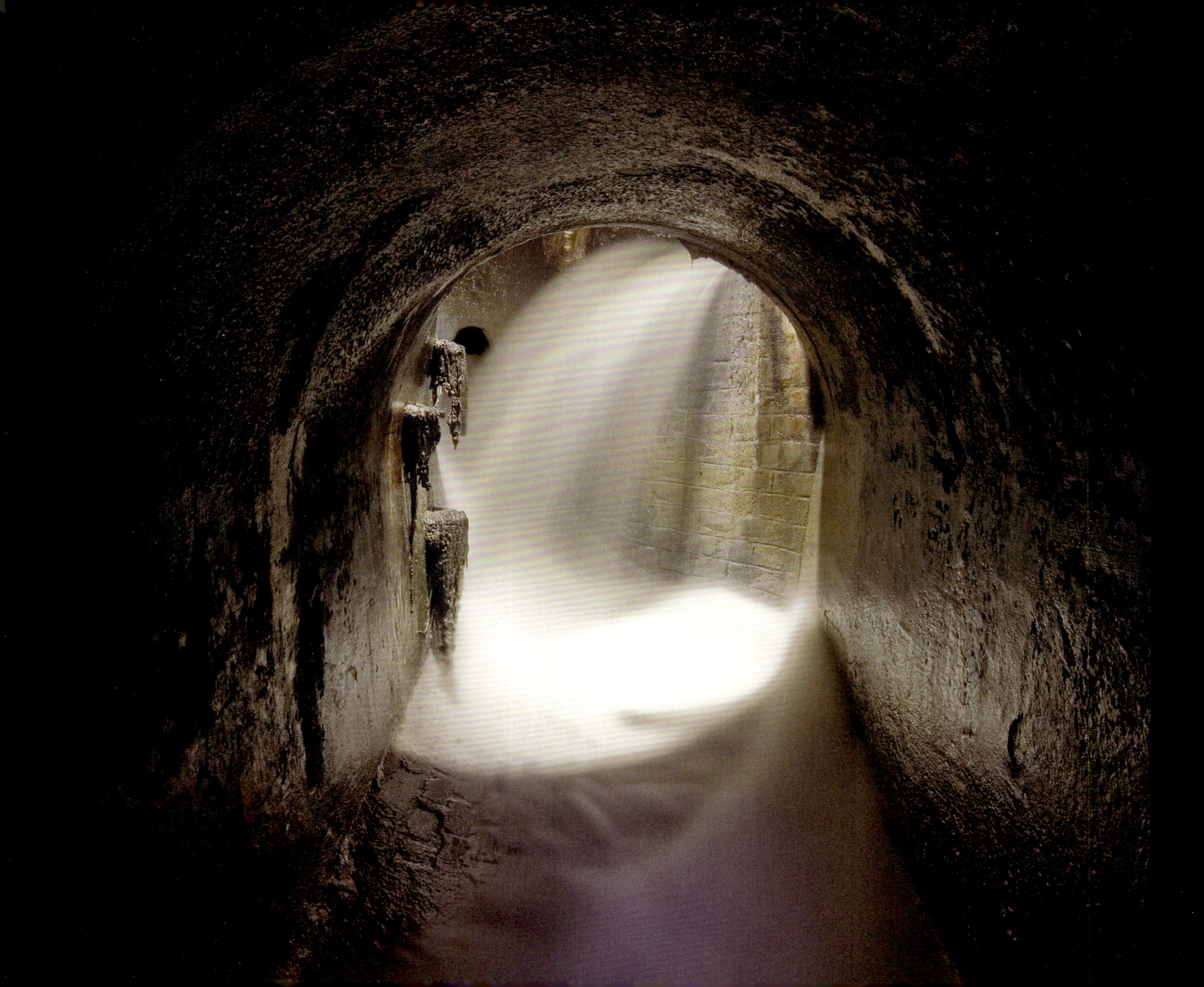

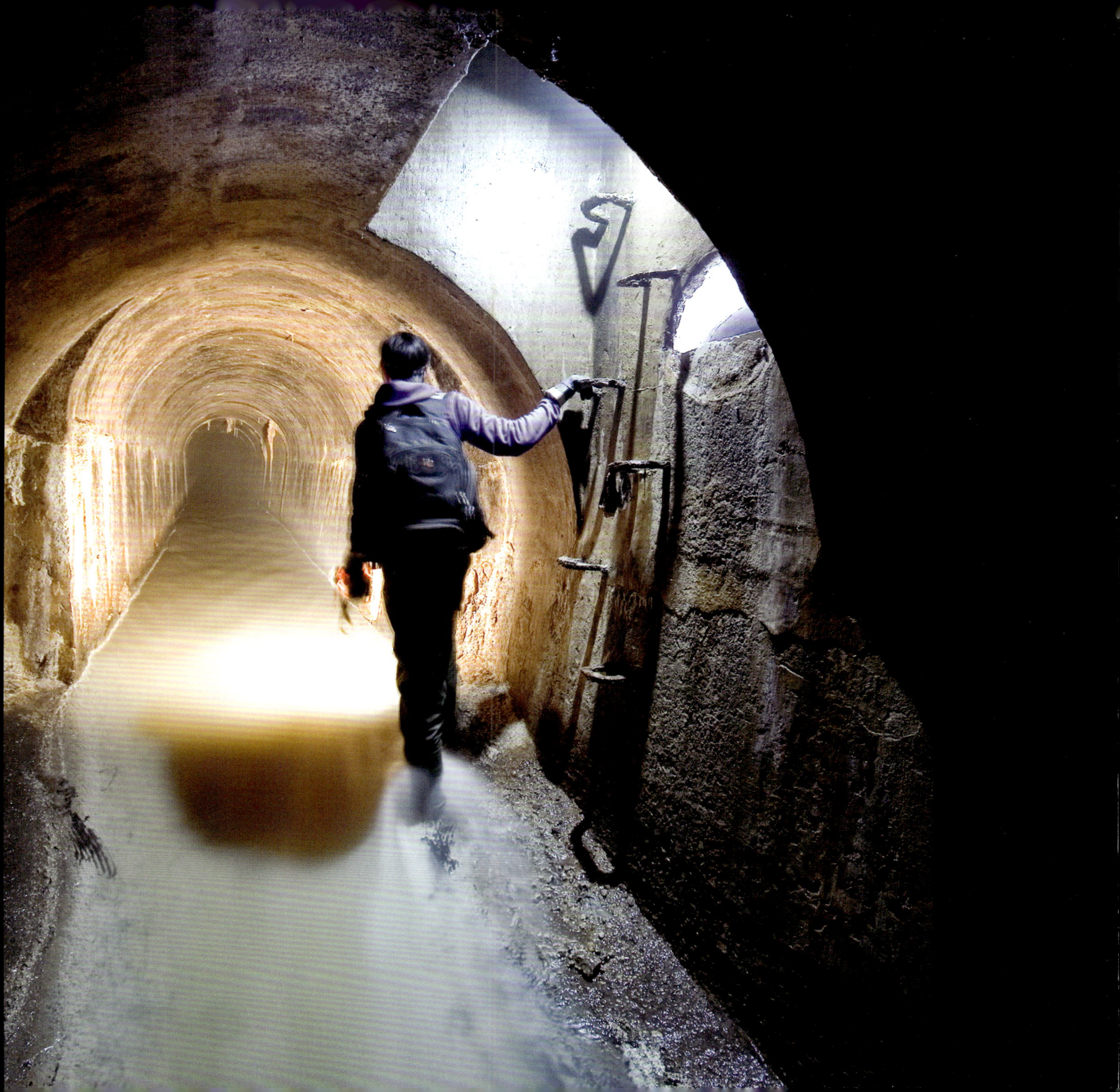

Voluminous

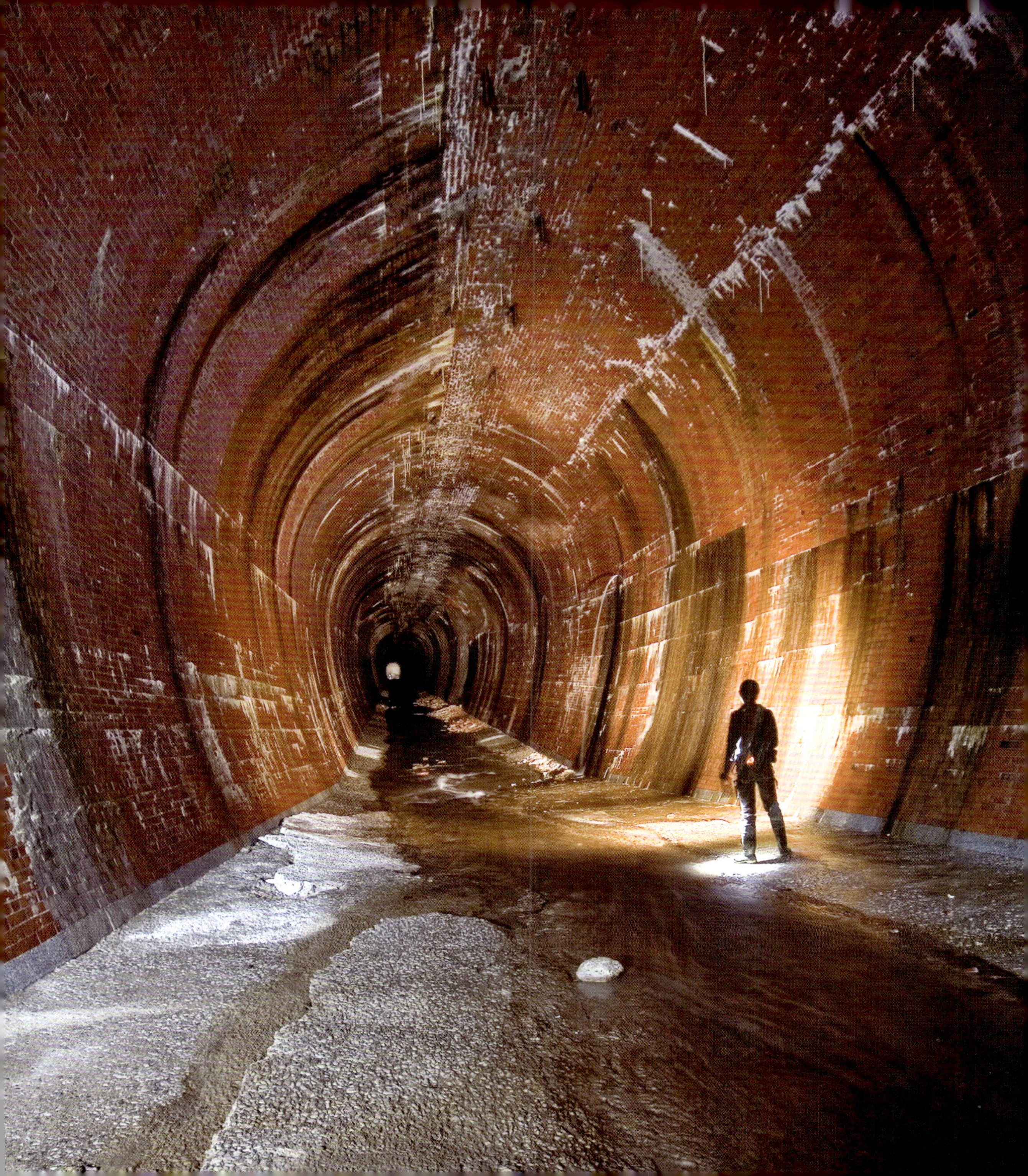

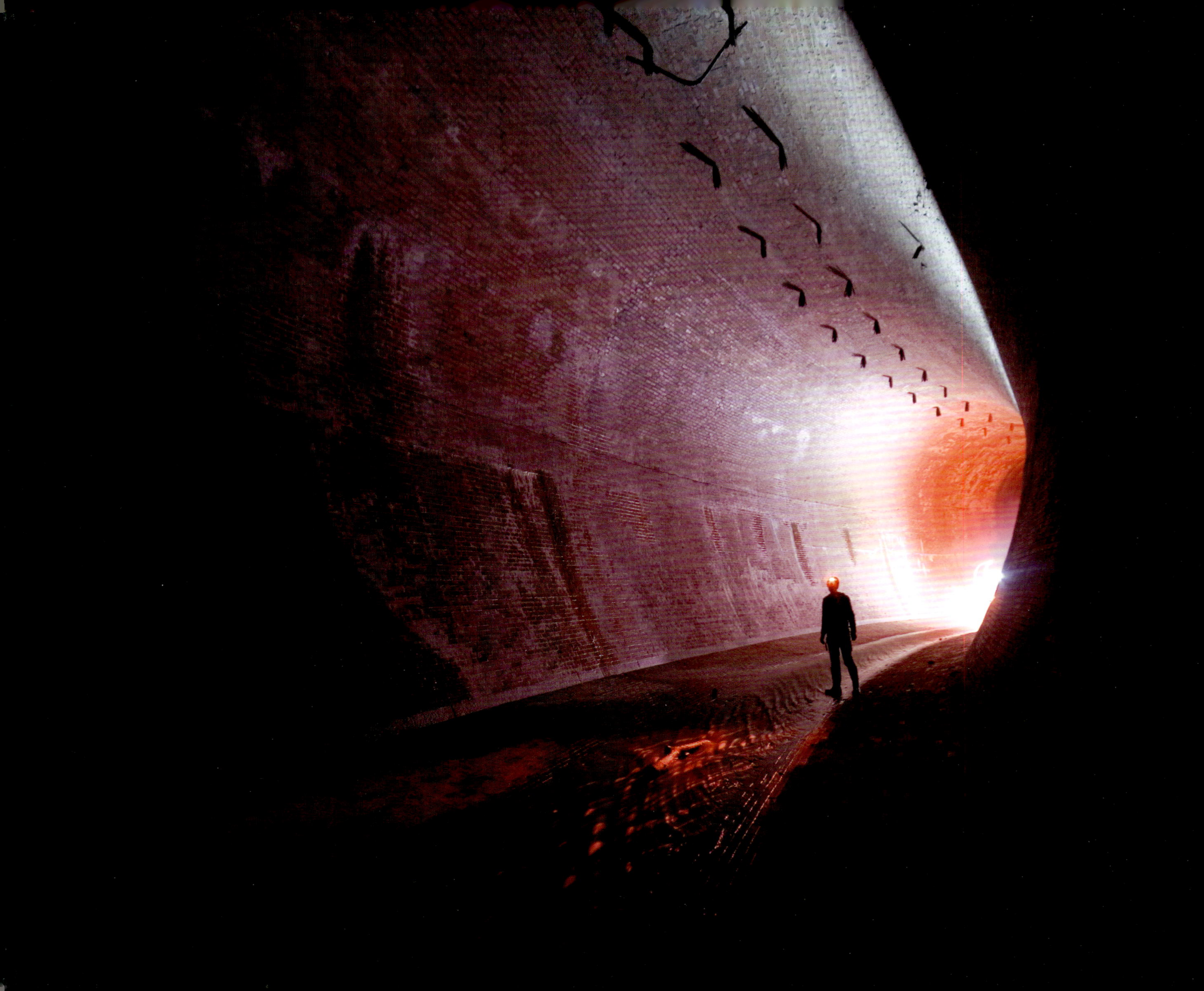

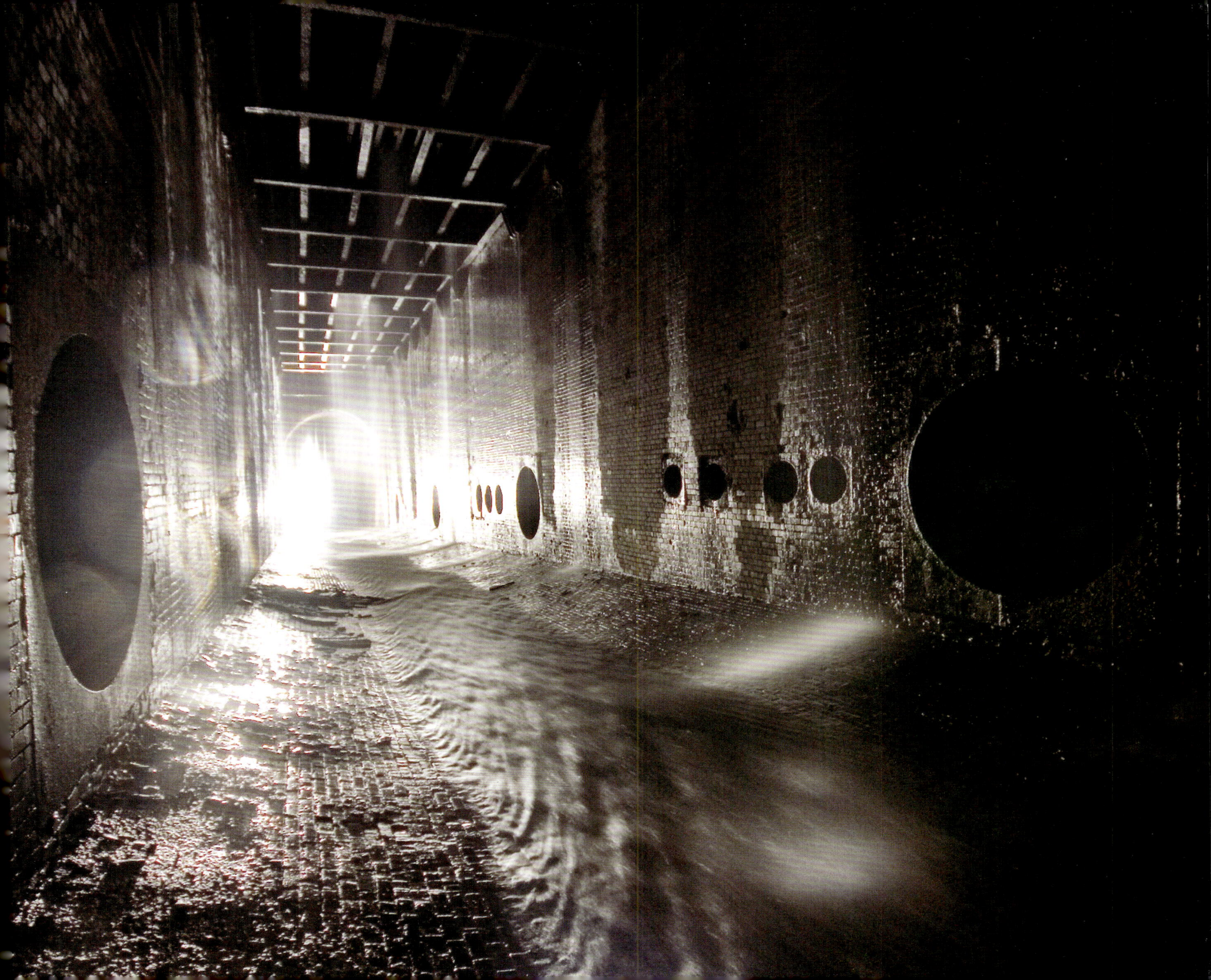

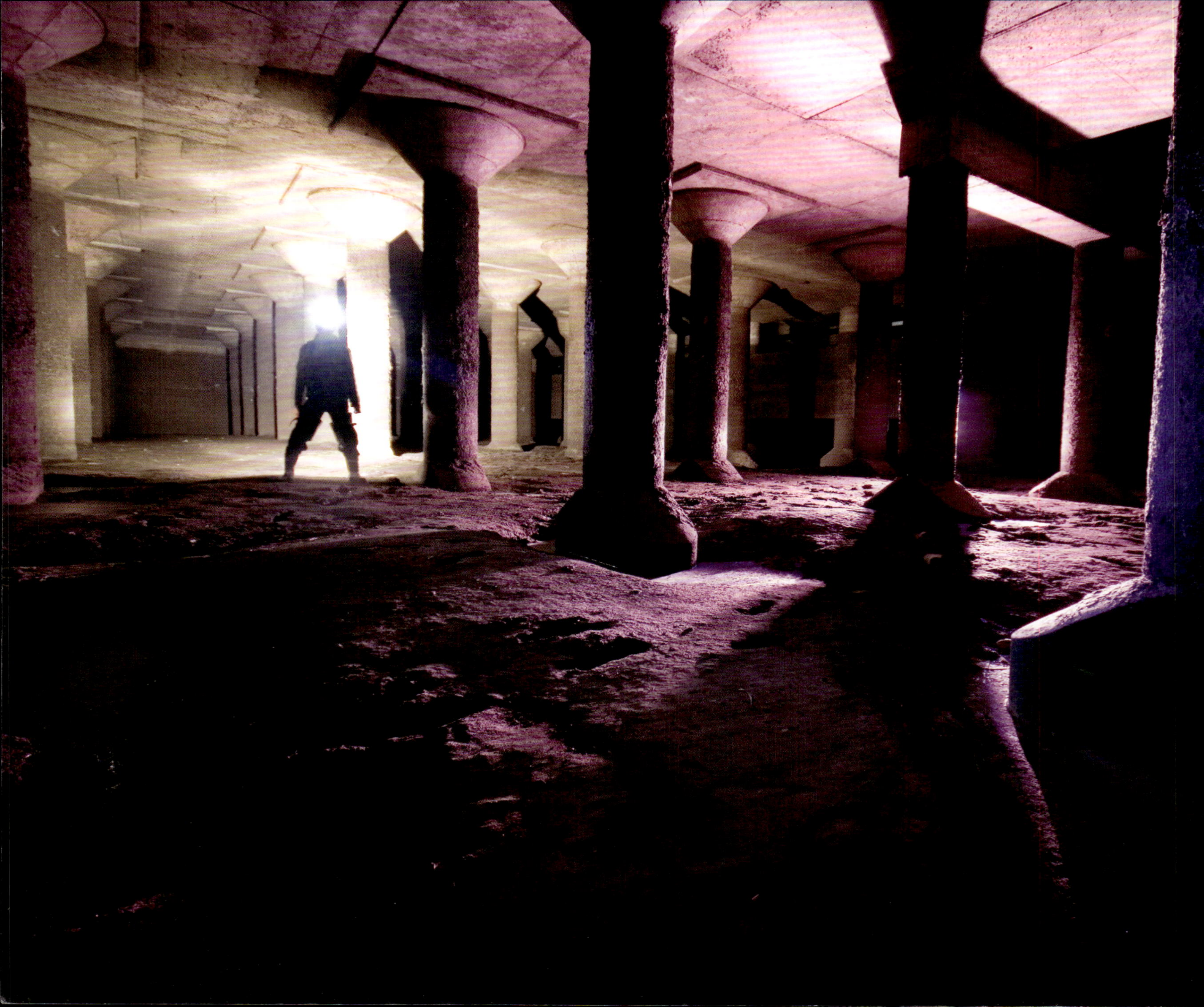

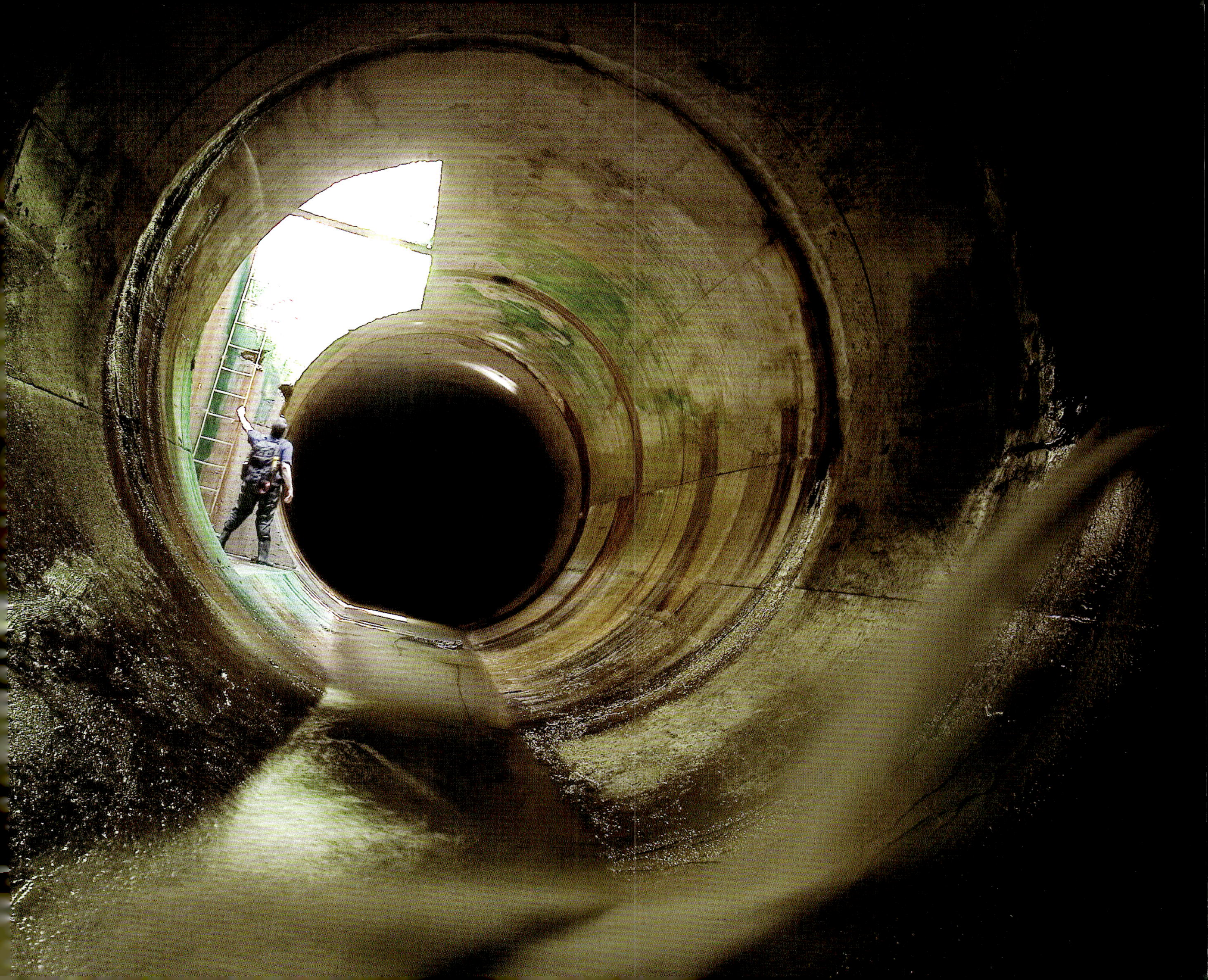

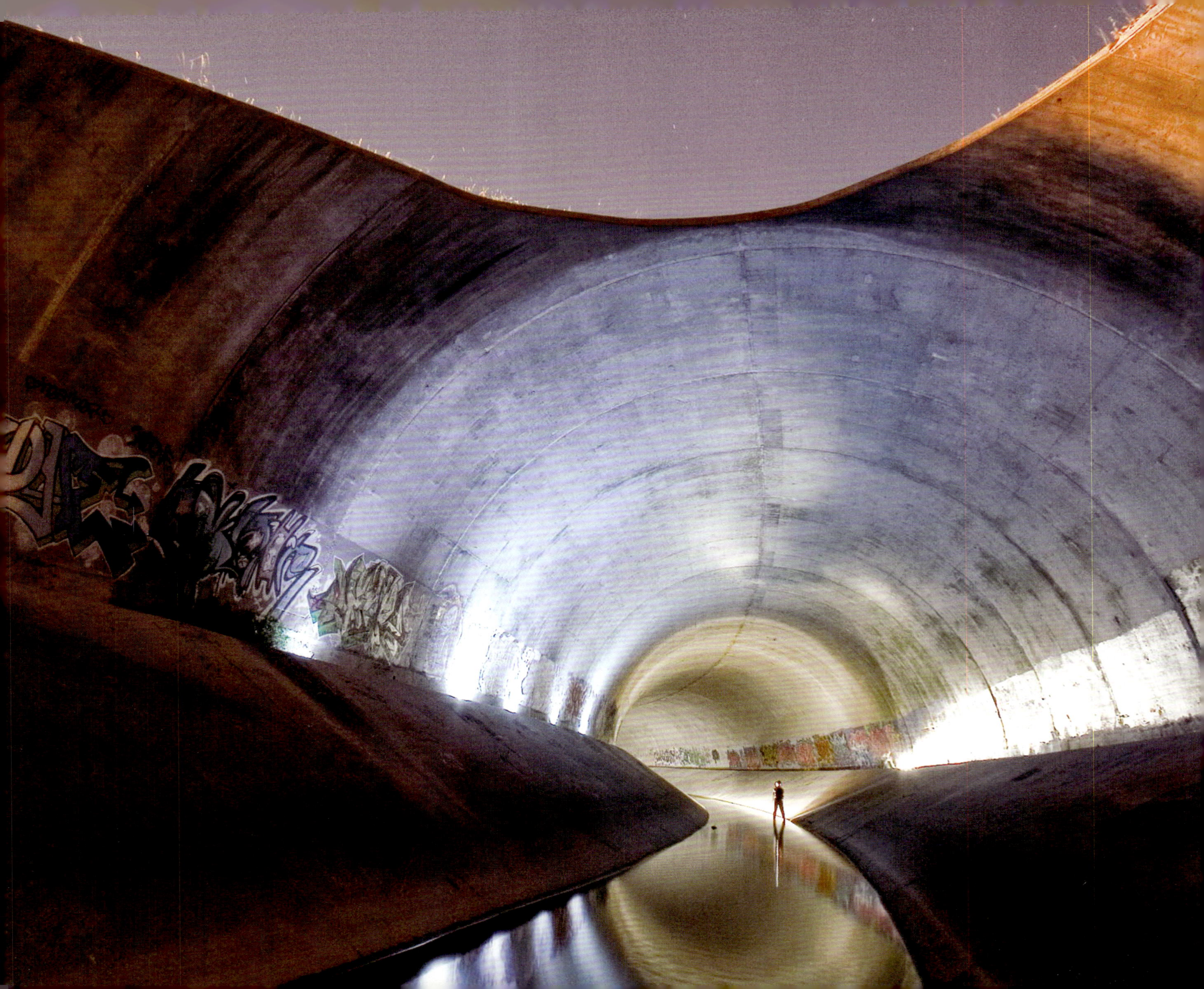

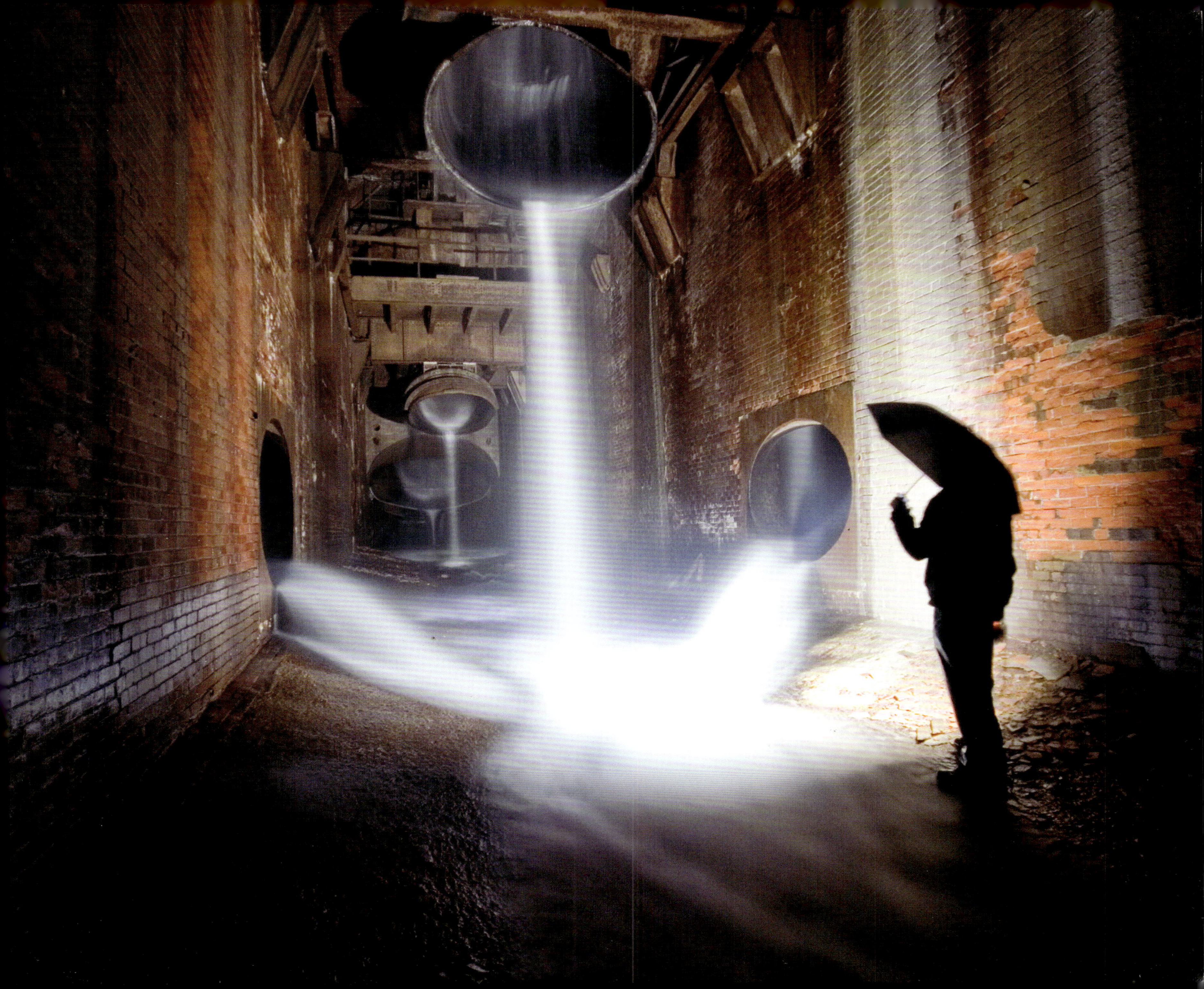

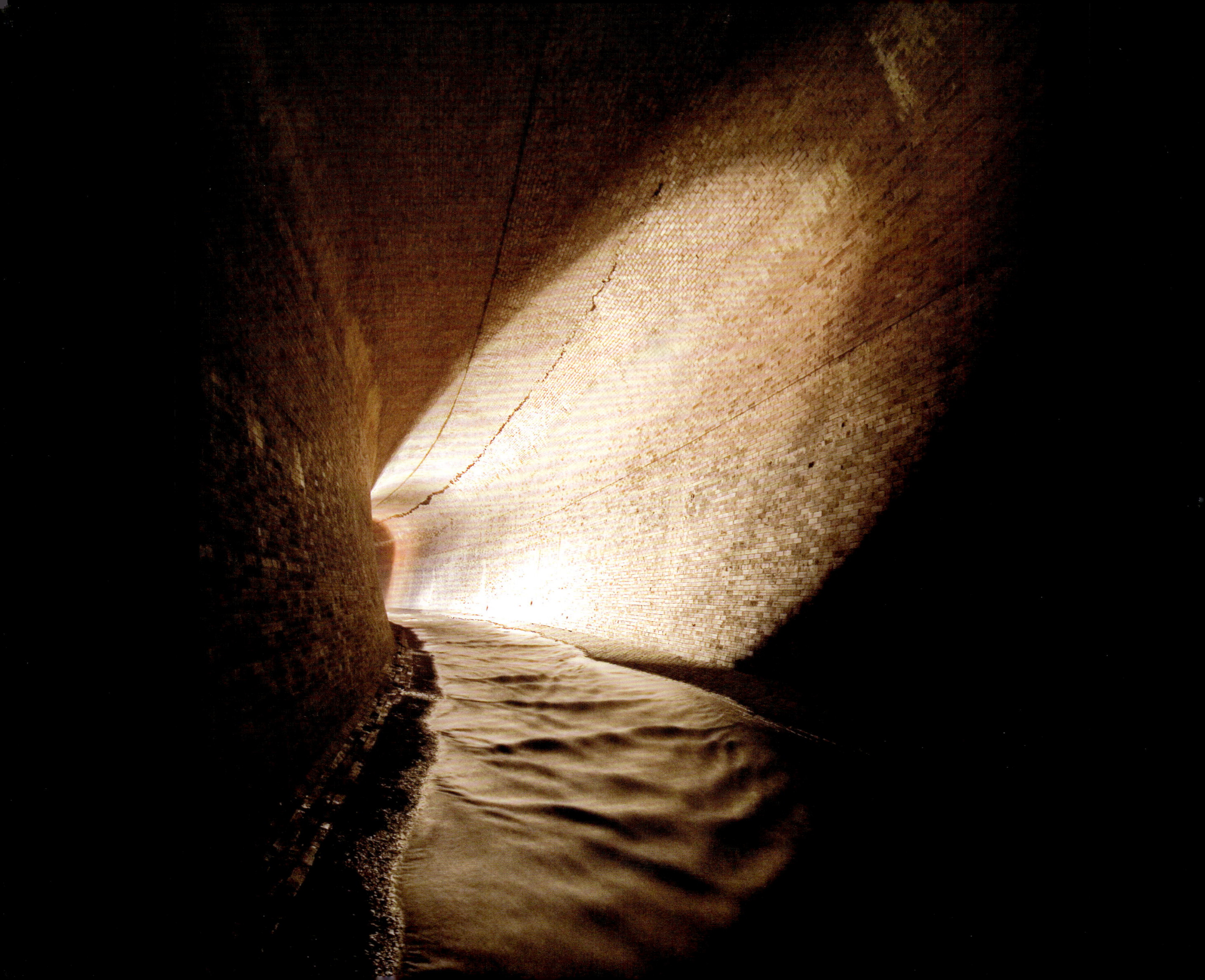

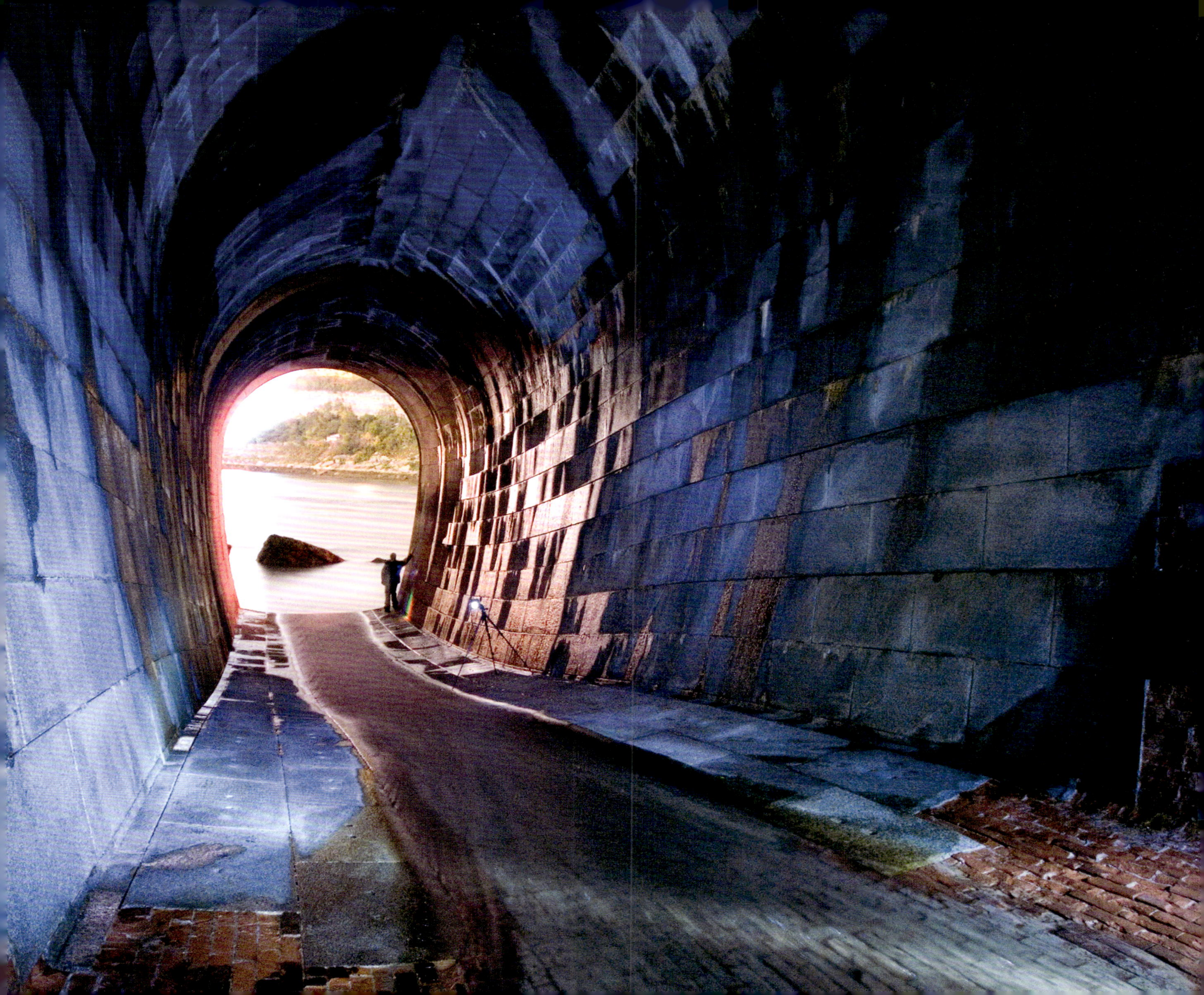

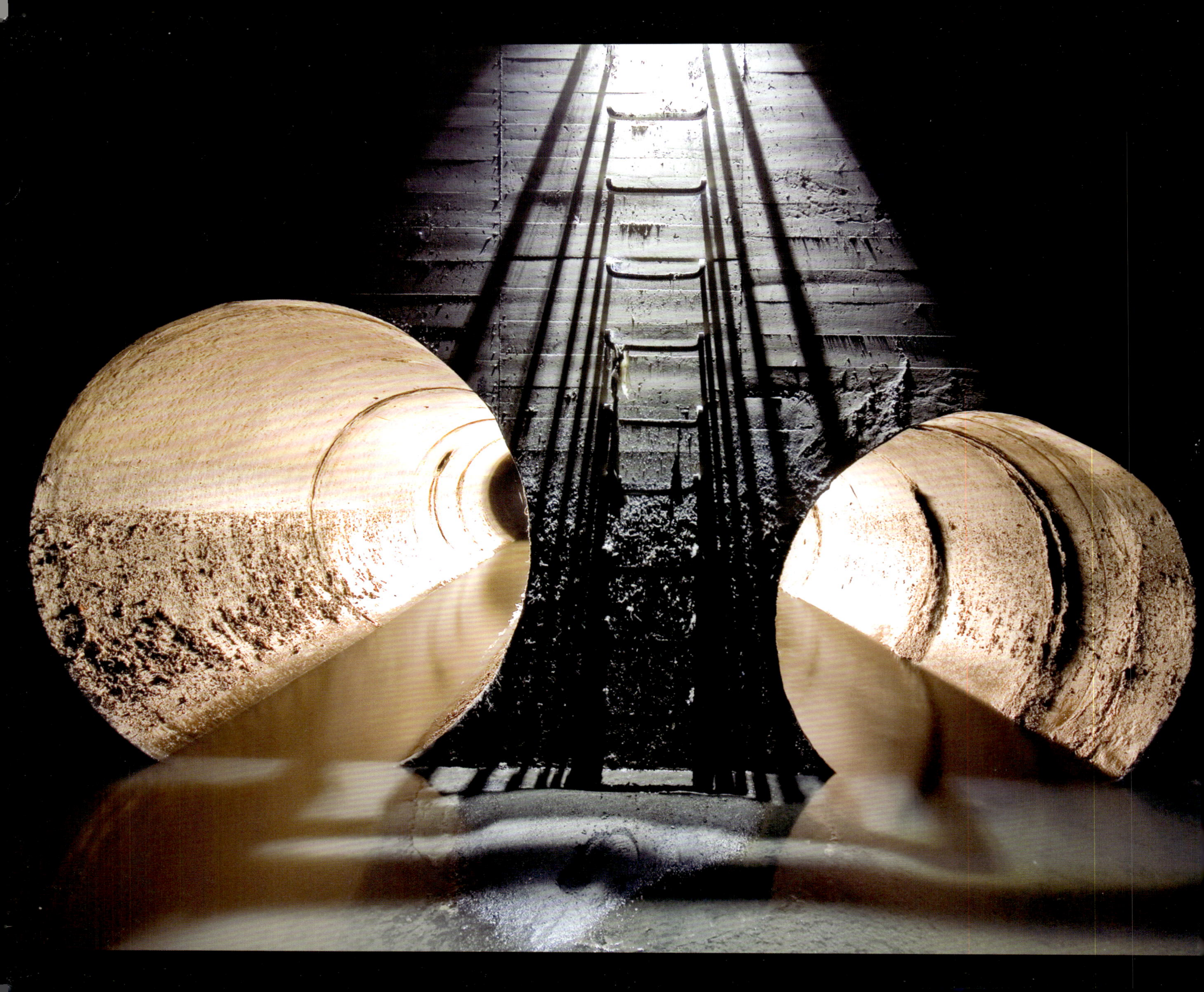